Introduction to Numerical Analysis and Applications

Introduction to Numerical Analysis and Applications

DONALD GREENSPAN
University of Wisconsin

MARKHAM PUBLISHING COMPANY
Chicago

MARKHAM MATHEMATICS SERIES
William J. LeVeque, Editor

Anderson, *Graph Theory and Finite Combinatorics*
Cantrell and Edwards, eds., *Topology of Manifolds*
Gaal, *Classical Galois Theory with Examples*
Gioia, *The Theory of Numbers: An Introduction*
Greenspan, *Introduction to Numerical Analysis and Applications*
Hu, *Calculus*
Hu, *Cohomology Theory*
Hu, *Elementary Functions and Coordinate Geometry*
Hu, *Linear Algebra*
Knopp, *Theory of Area*
Knopp, *Modular Functions in Analytic Number Theory*
Peterson, *Foundations of Algebra and Number Theory*
Stark, *An Introduction to Number Theory*

LECTURES IN ADVANCED MATHEMATICS

Davenport, *1. Multiplicative Number Theory*
Storer, *2. Cyclotomy and Difference Sets*
Engeler, *3. Formal Languages: Automata and Structures*
Garsia, *4. Topics in Almost Everywhere Convergence*

Printed in the United States of America
Library of Congress Catalog Card Number: 75–122306
Standard Book Number: 8410–1020–X

PREFACE

This text is designed for a one-semester, introductory course in the methods, theory, and applications of numerical analysis. Motivation for the selection and organization of topics is derived from scientific and technological applications of high-speed digital computers. The book attempts to excite and to maintain the students' interest by considering a broad spectrum of both classical and modern topics. Time and space limitations did require the deletion of a few important topics. I made decisions concerning such deletions in accordance with the increasing availability of full courses on special aspects of numerical analysis, and in accordance with the increasing needs and demands of scientists and engineers for numerical methods that apply to nonlinear dynamical problems.

I hope that study of this book will enable the student to quickly reach the point at which he can apply the methods and theory to problems significant to him.

For their special help in the preparation of the manuscript, I wish to thank Sandie Jones, Grace Krewson, David Schultz, and the alert, responsive students of my Computer Sciences 412 course.

D.G.

May, 1970
Madison, Wisconsin

CONTENTS

Chapter 1

APPROXIMATION

1.1. Introduction

In this book we shall study the application of modern high-speed digital computers (hereafter usually called only *computers*) to the solution of mathematical problems of physical origin. Although computers can store and retrieve numbers with great facility, attention will be directed only to using their enormous power to perform addition, subtraction, multiplication, and division. Thus we must develop and evaluate *arithmetic methods* for the study of both mathematical and physical problems.

More precisely, one defines a *computer algorithm* as a finite sequence of computer operations designed to yield an approximate solution of a given mathematical problem. *Numerical analysis* is that subject which is concerned with the development of computer algorithms and with the theory supporting the value of the algorithms. The development of an algorithm often requires extensive experimentation on a computer, whereas establishment of theoretical support always requires complete mathematical rigor.

1.2. Discrete Functions

In the laboratory, where all scientific endeavor begins, experimentation yields finite sets of data from which one tries to deduce meaningful relationships. Such finite sets of data are represented easily mathematically by *discrete functions*, which can be defined formally as follows.

For two real numbers a and b that satisfy $a < b$, let $[a,b]$ denote the interval of real numbers $a \leq x \leq b$. Subdivide $[a,b]$ into n equal parts, each of length $\Delta x = (b - a)/n$, by the $n + 1$ linearly ordered points $a = x_0 < x_1 < x_2 < \cdots < x_{n-1} < x_n = b$. Then the set $x_0, x_1, x_2, \ldots, x_n$ is called an R_{n+1} set. For example, the interval [0,2] is divided into five equal parts, each of length $\Delta x = \frac{2}{5}$, by the R_6 set $x_0 = 0$, $x_1 = \frac{2}{5}$, $x_2 = \frac{4}{5}$, $x_3 = \frac{6}{5}$, $x_4 = \frac{8}{5}$, $x_5 = 2$.

Definition 1.1. A function $y = f(x)$ that is defined only on an R_{n+1} set is called a *discrete function.*

The notation $y_i = f(x_i)$, in which x and y have the same subscripts, will be especially convenient when dealing with discrete functions. Note also that throughout, unless otherwise stated, *all* functions to be considered will be real-valued.

Example 1. Let [0,2] be divided into five equal parts by the R_6 set $x_0 = 0$, $x_1 = \frac{2}{5}$, $x_2 = \frac{4}{5}$, $x_3 = \frac{6}{5}$, $x_4 = \frac{8}{5}$, $x_5 = 2$. Then $y_i = x_i^2$, $i = 0, 1, 2, 3, 4, 5$, defines a discrete function on R_6. Also, $y_i = x_i^2$, $x_i \in R_6$, defines the same function, where, of course, "$x_i \in R_6$" is used in the customary mathematical way to mean "x_i is an element in the set R_6."

Example 2. Let [0,2] be divided into five equal parts by the R_6 set $x_0 = 0$, $x_1 = \frac{2}{5}$, $x_2 = \frac{4}{5}$, $x_3 = \frac{6}{5}$, $x_4 = \frac{8}{5}$, $x_5 = 2$. Then $f(x_0) = 0$, $f(x_1) = 1$, $f(x_2) = -1$, $f(x_3) = 0$, $f(x_4) = 2$, $f(x_5) = -2$ defines a discrete function on R_6.

Definition 1.2. If $y_i = f(x_i)$, $x_i \in R_{n+1}$, is a discrete function, then the set of points (x_i, y_i), $i = 0, 1, \ldots, n$ is called the *graph of the function.*

Example 1. Let [0,2] be divided into five equal parts by the R_6 set $x_0 = 0$, $x_1 = \frac{2}{5}$, $x_2 = \frac{4}{5}$, $x_3 = \frac{6}{5}$, $x_4 = \frac{8}{5}$, $x_5 = 2$. The graph of the discrete function $y_i = x_i^2$, $x_i \in R_6$, is shown in Figure 1.1.

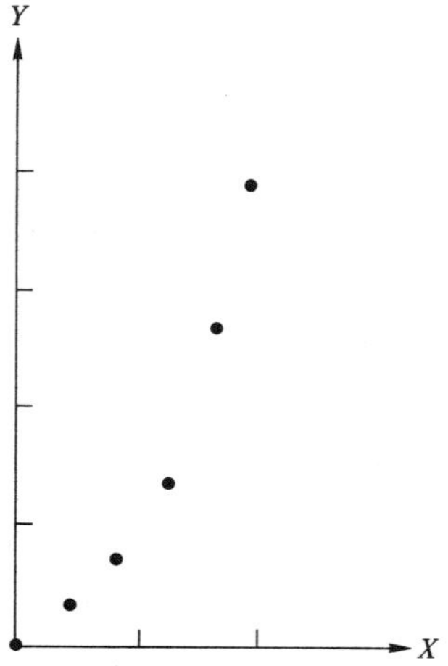

Figure 1.1

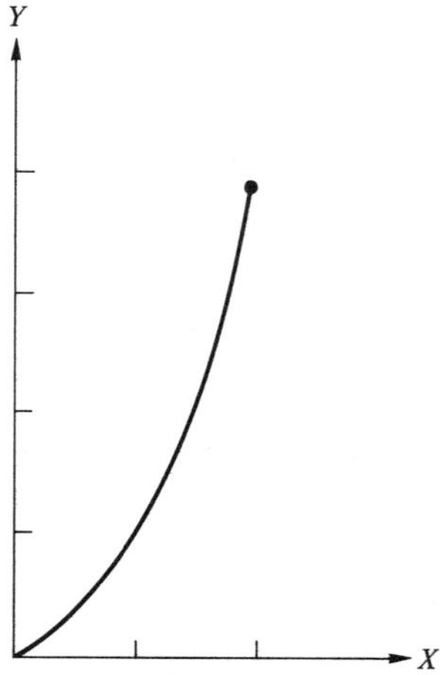

Figure 1.2

Example 2. Let [0,2] be divided into 10^5 equal parts by the $R_{100,001}$ set $x_i = 2i/10^5$, $i = 0, 1, 2, \ldots, 10^5$. The graph of the discrete function $y_i = x_i^2$, $x_i \in R_{100,001}$, is shown in Figure 1.2.

With regard to the last example, note that the graph of the discrete function appears to the naked eye to be identical with the graph of $y = x^2, 0 \le x \le 2$. The idea of "packing" a large, but finite, number of points sufficiently close so that each is physically indistinguishable from others nearby is, of course, the basis of television, and suggests that finite sets might serve one's purposes just as well, if not better, than infinite ones. We shall return to consider such a proposal in greater detail later. For the present we shall consider the classical problem of *interpolation.*

1.3. Interpolation

Interpolation is the process by which one approximates a discrete function, given on an R_{n+1} set $x_0, x_1, \ldots, x_n$, by a continuous function defined on $a \le x \le b$. There are many ways to interpolate and we shall study only some of those which are of broad, general use.

Definition 1.3. Let $y_i = f(x_i)$, $i = 0, 1, 2, \ldots, n$, be a discrete function that is defined on a given R_{n+1} set. Then the *piecewise linear interpolating function* $L(x)$ of y_i is the function whose graph consists of straight-line segments joining consecutive pairs of points (x_i, y_i), (x_{i+1}, y_{i+1}), for $i = 0, 1, 2, \ldots, n-1$.

Example. On the R_6 set $x_0 = 0$, $x_1 = 1$, $x_2 = 2$, $x_3 = 3$, $x_4 = 4$, $x_5 = 5$, let the discrete function $y_i = f(x_i)$, $i = 0, 1, 2, 3, 4, 5$, be defined by $y_0 = 0$, $y_1 = y_2 = y_3 = 1$, $y_4 = 3$, $y_5 = 2$. Then the piecewise linear interpolating function $L(x)$ of y_i is defined on $0 \le x \le 5$ by

$$L = \begin{cases} x, & 0 \le x \le 1, \\ 1, & 1 \le x \le 3, \\ 2x - 5, & 3 \le x \le 4, \\ -x + 7, & 4 \le x \le 5, \end{cases}$$

and its graph is given in Figure 1.3.

From a knowledge of the physics associated with a given set of data, one may decide that piecewise linear interpolation is not adequate. It may be more desirable to construct an interpolating function that is piecewise parabolic, and this can be done as follows.

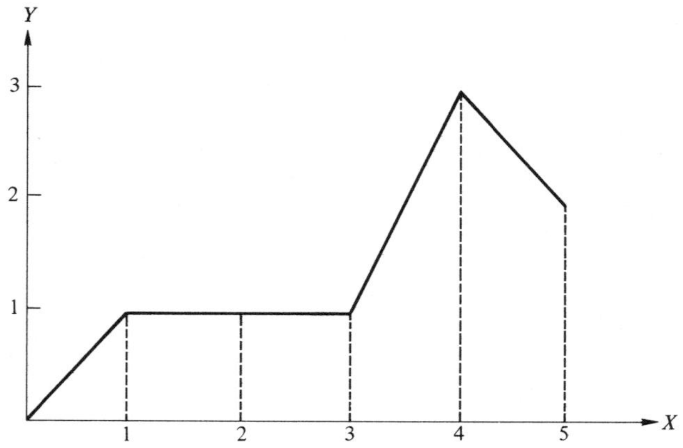

Figure 1.3

Consider first three points (x_0, y_0), (x_1, y_1), (x_2, y_2), where x_0, x_1, x_2 belong to an R_{n+1} set. If these points are to be on the graph of

$$y = a + bx + cx^2, \qquad x_0 \le x \le x_2, \tag{1.1}$$

then

$$\begin{aligned} y_0 &= a + bx_0 + cx_0^2, \\ y_1 &= a + bx_1 + cx_1^2, \\ y_2 &= a + bx_2 + cx_2^2, \end{aligned}$$

the unique solution of which, for a, b, and c, is

$$a = \frac{y_0(x_1x_2)(x_2 - x_1) + y_1(x_0x_2)(x_0 - x_2) + y_2(x_0x_1)(x_1 - x_0)}{(x_1 - x_0)(x_2 - x_0)(x_2 - x_1)}, \tag{1.2}$$

$$b = \frac{y_1(x_2^2 - x_0^2) + y_2(x_0^2 - x_1^2) + y_0(x_1^2 - x_2^2)}{(x_1 - x_0)(x_2 - x_0)(x_2 - x_1)}, \tag{1.3}$$

$$c = \frac{y_2(x_1 - x_0) + y_0(x_2 - x_1) + y_1(x_0 - x_2)}{(x_1 - x_0)(x_2 - x_0)(x_2 - x_1)}. \tag{1.4}$$

Note that in (1.1) we have allowed complete generality by not insisting that $c \neq 0$. Indeed, if (x_0, y_0), (x_1, y_1), and (x_2, y_2) are collinear, then $c = 0$ and (1.1) yields a desirable result. Implicit then in our discussion is that by "quadratic polynomial" we mean a polynomial whose degree is *at most* two.

From (1.1) through (1.4) we are led to the following definition.

Definition 1.4. For an R_{n+1} set $x_0, x_1, x_2, \ldots, x_n$, let n be even. Then the piecewise parabolic interpolating function $P(x)$ of a given discrete function $y_i = f(x_i)$, $x_i \in R_{n+1}$, is defined as follows. For each value of $i = 0, 1, 2, \ldots, (n/2) - 1$,

$$(1.5)\quad P(x) = a + bx + cx^2, \quad x_{2i} \le x \le x_{2i+2}, \qquad i = 0, 1, 2, \ldots, \frac{n}{2} - 1,$$

where

$$(1.6)\quad a = \frac{y_{2i}x_{2i+1}x_{2i+2}(x_{2i+2} - x_{2i+1}) + y_{2i+1}x_{2i}x_{2i+2}(x_{2i} - x_{2i+2}) + y_{2i+2}x_{2i}x_{2i+1}(x_{2i+1} - x_{2i})}{(x_{2i+1} - x_{2i})(x_{2i+2} - x_{2i})(x_{2i+2} - x_{2i+1})},$$

$$(1.7)\quad b = \frac{y_{2i+1}(x_{2i+2}^2 - x_{2i}^2) + y_{2i+2}(x_{2i}^2 - x_{2i+1}^2) + y_{2i}(x_{2i+1}^2 - x_{2i+2}^2)}{(x_{2i+1} - x_{2i})(x_{2i+2} - x_{2i})(x_{2i+2} - x_{2i+1})},$$

$$(1.8)\quad c = \frac{y_{2i+2}(x_{2i+1} - x_{2i}) + y_{2i}(x_{2i+2} - x_{2i+1}) + y_{2i+1}(x_{2i} - x_{2i+2})}{(x_{2i+1} - x_{2i})(x_{2i+2} - x_{2i})(x_{2i+2} - x_{2i+1})}.$$

Example. On the R_5 set $x_0 = 0$, $x_1 = 1$, $x_2 = 2$, $x_3 = 3$, $x_4 = 4$, let the discrete function $y_i = f(x_i)$ be defined by $y_0 = 0$, $y_1 = 1$, $y_2 = 4$, $y_3 = 4$, $y_4 = 2$. For $i = 0$ one has $x_{2i} = x_0 = 0$, $x_{2i+1} = x_1 = 1$, $x_{2i+2} = x_2 = 2$, $y_{2i} = y_0 = 0$, $y_{2i+1} = y_1 = 1$, $y_{2i+2} = y_2 = 4$, from which it follows that $a = 0$, $b = 0$, $c = 1$. Thus, from (1.5), one has

$$(1.9)\qquad P(x) = x^2, \qquad 0 \le x \le 2.$$

For $i = 1$ one has $x_{2i} = x_2 = 2$, $x_{2i+1} = x_3 = 3$, $x_{2i+2} = x_4 = 4$, $y_{2i} = y_2 = 4$, $y_{2i+1} = y_3 = 4$, $y_{2i+2} = y_4 = 2$, from which it follows that $a = -2$, $b = 5$, $c = -1$. Thus, from (1.5), one has

$$(1.10)\qquad P(x) = -2 + 5x - x^2, \qquad 2 \le x \le 4.$$

From (1.9) and (1.10) it follows finally that

$$P(x) = \begin{cases} x^2, & 0 \le x \le 2, \\ -2 + 5x - x^2, & 2 \le x \le 4 \end{cases}$$

is the piecewise parabolic interpolating function for the given discrete function. The graph of $P(x)$ is given in Figure 1.4.

1.4. Lagrange Interpolation Formulas

In Section 1.3 it was shown how to construct a piecewise linear function through consecutive pairs of points and a piecewise quadratic function

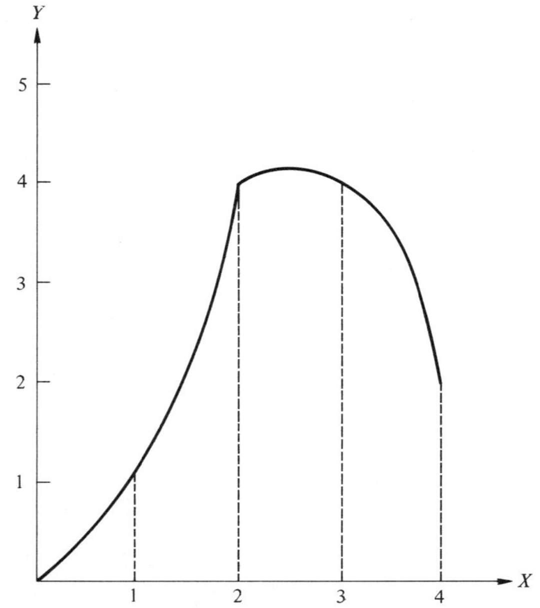

Figure 1.4

through consecutive triplets of points. One can proceed in this spirit to construct a piecewise cubic function through consecutive quadruplets of points, a piecewise quartic function through consecutive quintuplets of points, and so on. Instead of treating each such possibility in an individual manner, let us try to develop some intuition that will enable us to write down at once the desired result for any given number of points.

Substitution of (1.2) through (1.4) into (1.1) and recombination implies that

$$
\begin{aligned}
y = {} & \frac{1}{(x_2 - x_1)(x_2 - x_0)(x_1 - x_0)} \\
& \times \{y_0[x_1x_2^2 - x_2x_1^2 + xx_1^2 - xx_2^2 + x_2x^2 - x_1x^2] \\
& + y_1[x_2x_0^2 - x_0x_2^2 + xx_2^2 - xx_0^2 - x^2x_0 - x_2x^2] \\
& + y_2[x_0x_1^2 - x_1x_0^2 + xx_0^2 - xx_1^2 + x_1x^2 - x_0x^2]\} \\
= {} & \frac{1}{(x_2 - x_1)(x_2 - x_0)(x_1 - x_0)} \\
& \times \{y_0[x_2x^2 - x_1x^2 - x_2^2x + x_1x_2x - x_1x_2x + x_1^2x + x_1x_2^2 - x_1^2x_2] \\
& + y_1[x^2x_0 - xx_0x_2 - xx_0^2 + x_0^2x_2 - x_2x^2 + xx_2^2 + xx_0x_2 - x_0x_2^2] \\
& + y_2[x_1x^2 - xx_1^2 - xx_0x_1 + x_0x_1^2 - x^2x_0 + xx_0x_1 + xx_0^2 - x_0^2x_1]\}
\end{aligned}
$$

$$= \frac{1}{(x_2 - x_1)(x_2 - x_0)(x_1 - x_0)}[y_0(x - x_1)(x_2 - x_1)(x - x_2)$$
$$+ y_1(x_0 - x_2)(x - x_2)(x - x_0)$$
$$+ y_2(x_1 - x_0)(x - x_1)(x - x_0)],$$

so that, finally,

$$y = y_0\frac{(x - x_1)(x - x_2)}{(x_0 - x_1)(x_0 - x_2)} + y_1\frac{(x - x_2)(x - x_0)}{(x_1 - x_2)(x_1 - x_0)} + y_2\frac{(x - x_1)(x - x_0)}{(x_2 - x_1)(x_2 - x_0)}. \tag{1.11}$$

Of course (1.11) is identical with (1.1), where *a*, *b*, and *c* are determined by (1.2) through (1.4). But (1.11) has a much clearer structure than (1.1), for (1.11) consists of exactly three quadratic terms, each of which has a characterizable form. Consider, for example, the first term,

$$y_0\frac{(x - x_1)(x - x_2)}{(x_0 - x_1)(x_0 - x_2)}. \tag{1.12}$$

This term is associated with the first point (x_0,y_0) of the three points (x_0,y_0), (x_1,y_1), (x_2,y_2). The denominator of (1.12) is constructed from x_0 by subtracting from it all other possible choices of x_i and by multiplying the resulting differences. The numerator is constructed from the denominator by replacing x_0 by x. The coefficient y_0 is the value of y corresponding to x_0. In the same spirit, it follows that corresponding to (x_1,y_1) we would have a term whose denominator is $(x_1 - x_0)(x_1 - x_2)$, whose numerator is $(x - x_0) \times (x - x_2)$, and whose coefficient is y_1; corresponding to (x_2,y_2) one has a term whose denominator is $(x_2 - x_0)(x_2 - x_1)$, whose numerator is $(x - x_0)(x - x_1)$, and whose coefficient is y_2. The sum of the three resulting terms is, of course, (1.11).

Suppose now that one wishes to construct a cubic polynomial through the four points (x_0,y_0), (x_1,y_1), (x_2,y_2), (x_3,y_3), where x_0, x_1, x_2, x_3 belong to an R_{n+1} set. If one would wish to guess what such a polynomial would look like from the discussion above, the reasoning would be as follows. There must be four terms. The first term would correspond to (x_0,y_0). The denominator would be

$$(x_0 - x_1)(x_0 - x_2)(x_0 - x_3),$$

the numerator would be

$$(x - x_1)(x - x_2)(x - x_3),$$

and the coefficient would be y_0. Thus the first term would be

$$y_0 \frac{(x - x_1)(x - x_2)(x - x_3)}{(x_0 - x_1)(x_0 - x_2)(x_0 - x_3)}.$$

The second term would correspond to (x_1, y_1) and would be

$$y_1 \frac{(x - x_0)(x - x_2)(x - x_3)}{(x_1 - x_0)(x_1 - x_2)(x_1 - x_3)}.$$

The third term would correspond to (x_2, y_2) and would be

$$y_2 \frac{(x - x_0)(x - x_1)(x - x_3)}{(x_2 - x_0)(x_2 - x_1)(x_2 - x_3)}.$$

The fourth term would correspond to (x_3, y_3) and would be

$$y_3 \frac{(x - x_0)(x - x_1)(x - x_2)}{(x_3 - x_0)(x_3 - x_1)(x_3 - x_2)}.$$

Thus the guess would be

$$\begin{aligned} y = {} & y_0 \frac{(x - x_1)(x - x_2)(x - x_3)}{(x_0 - x_1)(x_0 - x_2)(x_0 - x_3)} + y_1 \frac{(x - x_0)(x - x_2)(x - x_3)}{(x_1 - x_0)(x_1 - x_2)(x_1 - x_3)} \\ & + y_2 \frac{(x - x_0)(x - x_1)(x - x_3)}{(x_2 - x_0)(x_2 - x_1)(x_2 - x_3)} + y_3 \frac{(x - x_0)(x - x_1)(x - x_2)}{(x_3 - x_0)(x_3 - x_1)(x_3 - x_2)}. \end{aligned} \tag{1.13}$$

The fact that (1.13) is actually of the desired form follows because each term is cubic and because substitution of $x = x_0, x_1, x_2$, and x_3 into (1.13) implies that $y = y_0, y_1, y_2$, and y_3, respectively. If one wishes for theoretical purposes to multiply out (1.13) and rewrite it in the form

$$y = a + bx + cx^2 + dx^3, \tag{1.14}$$

this can be done. But if one views both (1.13) and (1.14) from the high-speed-computer point of view, both are relatively simple and (1.13) is just as acceptable a form as (1.14). Indeed, this is often true even for theoretical purposes.

Example. Find the cubic polynomial whose graph contains the four points (0,1), (1,2), (2,0), (3, −2).

Solution. Setting $x_0 = 0$, $x_1 = 1$, $x_2 = 2$, $x_3 = 3$, $y_0 = 1$, $y_1 = 2$, $y_2 = 0$, $y_3 = -2$ and substituting into (1.13) yields

$$y = -\tfrac{1}{6}(x - 1)(x - 2)(x - 3) + x(x - 2)(x - 3) - \tfrac{1}{3}x(x - 1)(x - 2). \tag{1.15}$$

Although not necessary, multiplication and recombination of (1.15) yields the equivalent form

$$y = \frac{x^3}{2} - 3x^2 + \frac{7}{2}x + 1.$$

The special forms (1.11) and (1.13) are called *Lagrange forms* of the corresponding interpolating polynomials. The general Lagrange form will be given in the next section, where some fundamental theorems will be proved that support and extend the ideas developed thus far.

1.5. General Theory

In this section let $x_0, x_1, \ldots, x_n$ be an R_{n+1} set and consider as given the distinct points $(x_0,y_0), (x_1,y_1), \ldots, (x_n,y_n)$.

Theorem 1.1. There exists a unique polynomial, of degree less than or equal to n, whose graph contains the points $(x_0,y_0), (x_1,y_1), \ldots, (x_n,y_n)$.
Proof. Let

$$Q_j(x) = \frac{(x - x_0)(x - x_1)\cdots(x - x_{j-2})(x - x_{j-1})(x - x_{j+1}) \times (x - x_{j+2})\cdots(x - x_n)}{(x_j - x_0)(x_j - x_1)\cdots(x_j - x_{j-2})(x_j - x_{j-1})(x_j - x_{j+1}) \times (x_j - x_{j+2})\cdots(x_j - x_n)} \tag{1.16}$$

and define $p(x)$ by

$$p(x) = \sum_{i=0}^{n} y_i Q_i(x). \tag{1.17}$$

Then $p(x)$ is a polynomial of degree less than or equal to n, because each of its terms is of degree less than or equal to n. Moreover, $p(x_i) = y_i, i = 0,1,\ldots,n$. Thus there exists at least one polynomial of degree less than or equal to n whose graph contains the given points.

Suppose there exists a second polynomial $q(x)$ of degree less than or equal to n whose graph contains the given points. Then $p(x) - q(x)$ is a polynomial of degree less than or equal to n which has $n + 1$ zeros; they are $x_0, x_1, \ldots, x_n$. Thus $p(x) - q(x) \equiv 0$, or $p(x) \equiv q(x)$. Thus $p(x)$ is unique and the theorem is proved.

Definition 1.5. The form (1.17), where Q_i is defined by (1.16), is called the *Lagrange form* of the interpolating polynomial through the given points.

Theorem 1.1 implies that in each example in Section 1.4 the polynomial that was constructed was, indeed, the unique such polynomial.

Intimately related to the problem of constructing a polynomial whose graph contains the $n + 1$ points $(x_0,y_0), (x_1,y_1), \ldots, (x_n,y_n)$ is the problem of how well this polynomial approximates on $x_0 \le x \le x_n$ a given continuous function whose graph also contains the given points. Such problems are answered partially by the following theorem.

Theorem 1.2. Let $x_0, x_1, \ldots, x_n$ be an R_{n+1} set. On $x_0 \le x \le x_n$, let $f(x)$ be continuous and have continuous derivatives up to and including order $n + 1$. Define y_i, $i = 0, 1, \ldots, n$, by

$$y_i = f(x_i), \qquad i = 0, 1, \ldots, n. \tag{1.18}$$

Let $p(x)$ be defined by (1.17). Then for each x in $x_0 \le x \le x_n$ there exists a point ξ_x in $x_0 \le x \le x_n$ such that

$$f(x) - p(x) = \frac{1}{(n + 1)!} L(x) f^{(n+1)}(\xi_x), \tag{1.19}$$

where

$$L(x) = (x - x_0)(x - x_1)\cdots(x - x_n). \tag{1.20}$$

Proof. If $x = x_i$, $i = 0, 1, 2, \ldots, n$, then $f(x_i) - p(x_i) = 0$, $L(x_i) = 0$, and ξ_x is arbitrary. Suppose then that x is fixed, satisfies $x_0 \le x \le x_n$, and that $x \ne x_i$, $i = 0, 1, \ldots, n$. Consider the auxiliary function $\phi(t)$ defined by

$$\phi(t) = f(t) - p(t) - \frac{f(x) - p(x)}{L(x)} L(t).$$

Then

$$\begin{aligned} \phi(x_k) &= f(x_k) - p(x_k) - \frac{f(x) - p(x)}{L(x)} L(x_k), \qquad k = 0, 1, 2, \ldots, n \\ &= y_k - y_k \\ &= 0, \end{aligned}$$

$$\phi(x) = f(x) - p(x) - \frac{f(x) - p(x)}{L(x)} L(x) \equiv 0. \tag{1.21}$$

Thus $\phi(t)$ has at least $n + 2$ distinct zeros in $x_0 \le t \le x_n$; they are $x_0, x_1, x_2, \ldots, x_n$ and x. Differentiation of $\phi(t)$ with respect to t would imply then, by Rolle's theorem, that ϕ' has at least $n + 1$ distinct zeros in $x_0 \le t \le x_n$, ϕ'' has at least n distinct zeros in $x_0 \le t \le x_n$, ϕ''' has at least

$n-1$ distinct zeros in $x_0 \le t \le x_n, \ldots$, and $\phi^{(n+1)}$ has at least one zero in $x_0 \le t \le x_n$. Let ξ_x be such a zero of $\phi^{(n+1)}$. Differentiating and setting $t = \xi_x$ implies that

$$\phi^{(n+1)}(\xi_x) = f^{(n+1)}(\xi_x) - p^{(n+1)}(\xi_x) - \frac{f(x) - p(x)}{L(x)} L^{(n+1)}(\xi_x). \tag{1.22}$$

But p is a polynomial of degree n, and L is a polynomial of degree $n-1$ with leading coefficient unity, so that $p^{(n+1)} = 0, L^{(n+1)} = (n+1)!$ Thus (1.22) implies that

$$\phi^{(n+1)}(\xi_x) = f^{(n+1)}(\xi_x) - \frac{f(x) - p(x)}{L(x)}(n+1)! = 0,$$

which, with (1.21), implies (1.19), and the theorem is proved.

Expression (1.19) is called the error in approximating $f(x)$ by $p(x)$.

Example. On $[0,\pi]$ let $f(x) = \sin x$. On the R_3 set $x_0 = 0$, $x_1 = \pi/2$, $x_2 = \pi$, define $y_0 = \sin x_0 = 0$, $y_1 = \sin x_1 = 1$, $y_2 = \sin x_2 = 0$. Let $p(x)$ be the quadratic interpolating polynomial through (x_0,y_0), (x_1,y_1), (x_2,y_2), so that

$$p(x) \equiv \frac{4}{\pi}x - \frac{4}{\pi^2}x^2. \tag{1.23}$$

The graphs of $f(x)$ and $p(x)$ are shown in Figure 1.5. For any value of x in $0 \le x \le \pi$, the expression (1.19) tells by how much $p(x)$ differs from $f(x)$

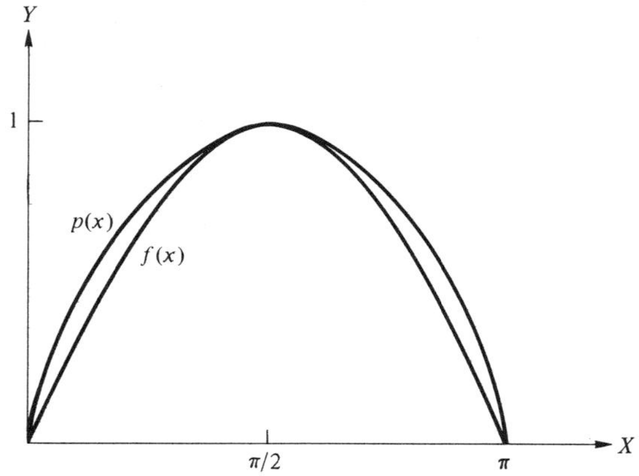

Figure 1.5

and in this case takes the form

$$\sin x - \left(\frac{4}{\pi}x - \frac{4}{\pi^2}x^2\right) = -\frac{1}{3!}(x)\left(x - \frac{\pi}{2}\right)(x - \pi)\cos \xi_x. \tag{1.24}$$

From (1.24) it follows that (1.23) does not differ from $\sin x$ at $x = 0, \pi/2, \pi$, whereas, at a point such as $\pi/6$, (1.23) differs from $\sin x$ by

$$\sin\frac{\pi}{6} - p\left(\frac{\pi}{6}\right) = -\frac{5\pi^3}{648}\cos \xi. \tag{1.25}$$

Although ξ is indeterminate, one does know that $|\cos \xi| \leq 1$, so it follows readily from (1.25) that

$$\left|\sin\frac{\pi}{6} - p\left(\frac{\pi}{6}\right)\right| \leq \frac{5\pi^3}{648}.$$

1.6. Least-Squares Interpolation

Thus far we have concerned ourselves with interpolating functions whose graphs contained *all* the points of a discrete function. But as discrete functions are modeled on experimental data and experimental data are always approximate in nature, we seek next to construct interpolating functions whose graphs, although not *containing* the points of a given discrete function, "fit the given function very closely." One such method is called the method of *least squares* and is described quite generally as follows.

Let $y_i = f(x_i)$, $i = 0, 1, \ldots, n$, be a given discrete function defined on an R_{n+1} set $x_0, x_1, \ldots, x_n$. Let $g_1(x), g_2(x), \ldots, g_m(x)$ be m functions, each defined and continuous on $x_0 \leq x \leq x_n$ and such that no one $g_i(x)$ is a linear combination of the others. The choice of these functions, in any particular problem, is usually a reasoned choice founded on an intimate knowledge of the physics of the problem. One wishes to construct a function

$$G(x) = a_1 g_1(x) + a_2 g_2(x) + \cdots + a_m g_m(x), \tag{1.26}$$

where $a_1, a_2, \ldots, a_m$ are constants, which "fits the data closely," which we interpret here to mean that

$$S = [G(x_0) - y_0]^2 + [G(x_1) - y_1]^2 + \cdots + [G(x_n) - y_n]^2 \tag{1.27}$$

is an absolute minimum. We minimize S by solving the system

$$\frac{\partial S}{\partial a_1} = 0, \frac{\partial S}{\partial a_2} = 0, \ldots, \frac{\partial S}{\partial a_m} = 0 \tag{1.28}$$

for $a_1, a_2, \ldots, a_m$. Substitution of these values into (1.26) yields the desired function G.

Example 1. In a certain experiment it is found that a particle's position y is $-1, 1, 3, 4$ at the respective times $x = 0, 1, 2, 3$. It is also observed that y appears to depend linearly on x. To fit the given data by a linear function, we set $m = 2$, $g_1(x) = 1$, and $g_2(x) = x$ and consider

$$G(x) = a_1 + a_2 x. \tag{1.29}$$

Then

$$S = [-1 - a_1]^2 + [1 - (a_1 + a_2)]^2 + [3 - (a_1 + 2a_2)]^2 + [4 - (a_1 + 3a_2)]^2$$

and system (1.28) reduces to

$$4a_1 + 6a_2 = 7,$$
$$6a_1 + 14a_2 = 19.$$

The solution of this system is $a_1 = -0.8$, $a_2 = 1.7$, and the desired least-squares fit (1.29) is

$$G(x) = -0.8 + 1.7x. \tag{1.30}$$

The graphs of (1.30) and the discrete function it fits are given in Figure 1.6.

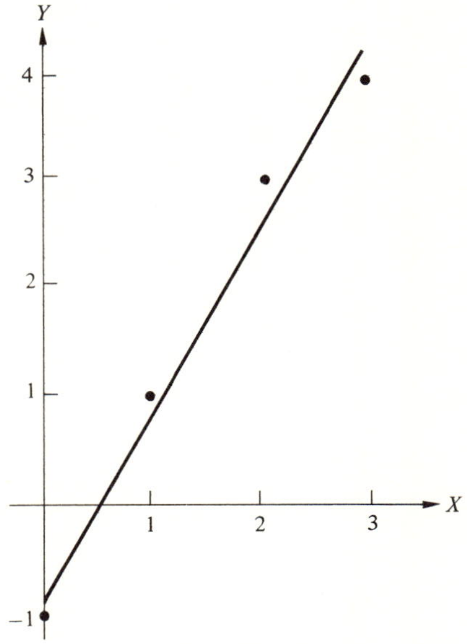

Figure 1.6

Example 2. In observing the motion of a particle in a plane, it is found that when $x = -3, -2, -1, 0, 1, 2, 3$, the respective y coordinates are $-0.71, -0.01, 0.51, 0.82, 0.88, 0.81, 0.49$. The graph of the resulting discrete function suggests that the particle's behavior may be quadratic in nature. Assuming this to be valid, we select $m = 3$, $g_1(x) = 1$, $g_2(x) = x$, and $g_3(x) = x^2$ and fit the data by least squares. Thus

$$G(x) = a_1 + a_2x + a_3x^2, \tag{1.31}$$

$$\begin{aligned} S = {} & [a_1 - 3a_2 + 9a_3 + 0.71]^2 + [a_1 - 2a_2 + 4a_3 + 0.01]^2 \\ & + [a_1 - a_2 + a_3 - 0.51]^2 + [a_1 - 0.82]^2 + [a_1 + a_2 + a_3 - 0.88]^2 \\ & + [a_1 + 2a_2 + 4a_3 - 0.81]^2 + [a_1 + 3a_2 + 9a_3 - 0.49]^2, \end{aligned}$$

and system (1.28) takes the form

$$\begin{aligned} 7a_1 + 28a_3 &= 2.79, \\ 28a_2 &= 5.61, \\ 28a_1 + 196a_3 &= 2.61. \end{aligned}$$

The solution of this system to three decimal places is $a_1 = 0.806$, $a_2 = 0.200$, $a_3 = -0.102$, so (1.31) takes the form

$$G(x) = 0.806 + 0.200x - 0.102x^2. \tag{1.32}$$

Both the given data and (1.32) are shown in Figure 1.7.

Example 3. On an oscilloscope, a certain periodic behavior is observed in the graphical form shown in Figure 1.8. To fit the given data graphically, it is decided that the functions $g_1(x), g_2(x), \ldots, g_m(x)$ should be periodic functions. Refined measurements indicate that when $x = 0, \pi/4, \pi/2, 3\pi/4, \pi$, the corresponding y values are $y = -0.9, 1.5, 3.1, 3, 1.1$. For the selections

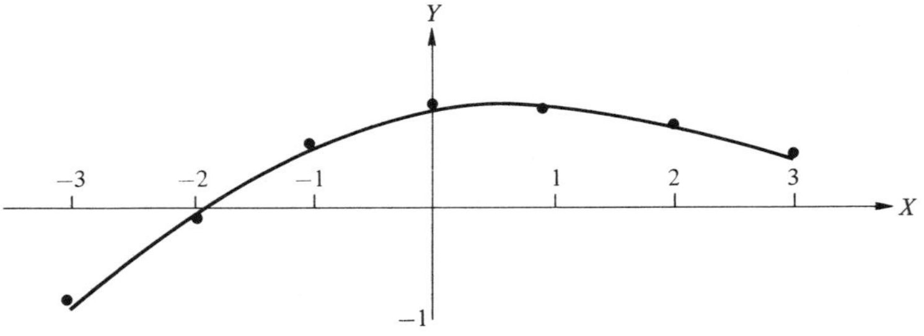

Figure 1.7

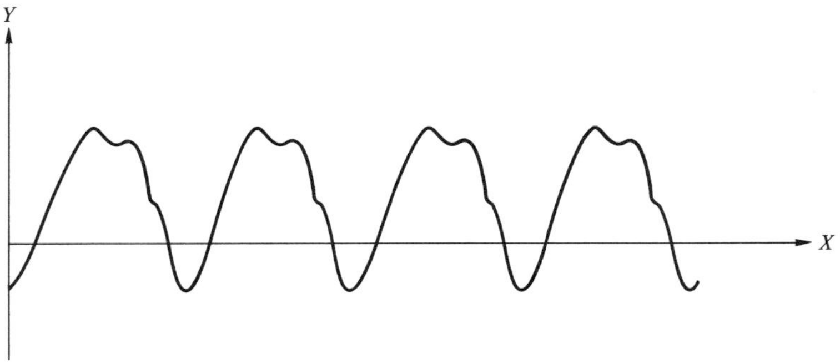

Figure 1.8

$m = 2$, $g_1(x) = \sin x$, and $g_2(x) = \cos x$, the function $G(x)$ is

$$G(x) = a_1 \sin x + a_2 \cos x.$$

System (1.28) becomes

$$2a_1 = \frac{4.5}{2}\sqrt{2} + 3.1,$$

$$3a_2 = -2 - \frac{1.5}{2}\sqrt{2},$$

from which one has $a_1 = 3.14$, $a_2 = -1.02$, and $G(x)$ is

$$G(x) = 3.14 \sin x - 1.02 \cos x.$$

1.7. Selected Theory for Least-Squares Interpolation

In this section we shall show theoretically that least-squares interpolation has a mathematical basis which makes it a reasonable numerical method. In order not to go too far afield, the discussion will be restricted to interpolation by polynomials.

Theorem 1.3. Let $x_0, x_1, \ldots, x_n$ be an R_{n+1} set and let $y_i = f(x_i)$ be a discrete function defined on R_{n+1}. Then for each fixed m that is less than $n + 1$, the polynomial

$$G(x) = a_1 + a_2 x + a_3 x^2 + \cdots + a_{m+1} x^m, \tag{1.33}$$

whose coefficients $a_1, a_2, \ldots, a_{m+1}$ are defined by system (1.28), is unique.

Proof. Since

$$S = \sum_{i=0}^{n} [G(x_i) - y_i]^2, \tag{1.34}$$

it follows from (1.33) and (1.34) that system (1.28) takes the form

$$\frac{\partial S}{\partial a_k} = 2 \sum_{i=0}^{n} [a_1 + a_2 x_i + a_3 x_i^2 + \cdots + a_{m+1} x_i^m - y_i] x_i^{k-1} = 0, \tag{1.35}$$

$$k = 1, 2, \ldots, m + 1.$$

From (1.35),

$$\sum_{i=0}^{n} y_i x_i^{k-1} = a_1 \sum_{i=0}^{n} x_i^{k-1} + a_2 \sum_{i=0}^{n} x_i^k + a_3 \sum_{i=0}^{n} x_i^{k+1} + \cdots + a_{m+1} \sum_{i=0}^{n} x_i^{k+m-1} = \sum_{j=1}^{m+1} \left[a_j \sum_{i=0}^{n} x_i^{k+j-2} \right]. \tag{1.36}$$

For simplicity, set

$$\sum_{i=0}^{n} x_i^k = S_k, \qquad \sum_{i=0}^{n} y_i x_i^k = T_k. \tag{1.37}$$

Then (1.36) implies that

$$\sum_{j=1}^{m+1} a_j S_{k+j-2} = T_{k-1}, \qquad k = 1, 2, \ldots, m + 1. \tag{1.38}$$

Now, (1.38) is a system of $(m + 1)$ linear algebraic equations in the $(m + 1)$ unknowns $a_1, a_2, \ldots, a_{m+1}$. Let the determinant of this system be Δ; that is, let

$$\Delta = |S_{k+j-2}|.$$

If $\Delta \neq 0$, then the theorem is valid. Consider the case $\Delta = 0$, and assume that the homogeneous system

$$\sum_{j=1}^{m+1} a_j S_{k+j-2} = 0, \qquad k = 1, 2, \ldots, m + 1 \tag{1.39}$$

has a *nonzero* solution vector $(a_1, a_2, \ldots, a_{m+1})$. Multiplication of the kth equation of system (1.39) by a_k and summing from $k = 1$ to $k = m + 1$

implies that

$$0 = \sum_{k=1}^{m+1} \left[a_k \left(\sum_{j=1}^{m+1} a_j S_{k+j-2} \right) \right] = \sum_{k=1}^{m+1} \left(a_k \left\{ \sum_{j=1}^{m+1} \left[a_j \left(\sum_{i=0}^{n} x_i^{j+k-2} \right) \right] \right\} \right)$$

$$= \sum_{k=1}^{m+1} \left(a_k \left\{ \sum_{j=1}^{m+1} \left[a_j \left(\sum_{i=0}^{n} x_i^{j-1} x_i^{k-1} \right) \right] \right\} \right)$$

$$= \sum_{i=0}^{n} \left[\left(\sum_{k=1}^{m+1} a_k x_i^{k-1} \right) \left(\sum_{j=1}^{m+1} a_j x_i^{j-1} \right) \right]$$

$$= \sum_{i=0}^{n} \left[\left(\sum_{j=1}^{m+1} a_j x_i^{j-1} \right) \left(\sum_{j=1}^{m+1} a_j x_i^{j-1} \right) \right] = \sum_{i=0}^{n} \left[\left(\sum_{j=1}^{m+1} a_j x_i^{j-1} \right) \right]^2$$

$$= \sum_{i=0}^{n} [G(x_i)]^2.$$

Hence

$$G(x_i) = 0, \qquad i = 0, 1, \ldots, n.$$

But $m < n + 1$, so (1.33) has more than m zeros and is therefore identically zero. Hence $(a_1, a_2, \ldots, a_{m+1})$ cannot be a *nonzero* vector, and the theorem follows readily by contradiction.

It is worth noting immediately that the particular case $m = n$ implies Theorem 1.1.

Theorem 1.3 assures one that (1.33) exists and is unique. But for additional confidence in least squares, it would be most important to know that $G(x)$ actually *minimizes*, and does not maximize or merely extremize, (1.34). That this is valid can be proved in complete generality, but rather than develop the advanced concepts needed for such a theorem, we shall prove a *very simple* form of the theorem in a way that can be extended readily to the general case.

Theorem 1.4. Let x_0, x_1, x_2 be an R_3 set and let $y_i = f(x_i)$ be a discrete function defined on R_3. Then the polynomial

$$G(x) = a_1 + a_2 x, \tag{1.40}$$

whose coefficients are determined from (1.28) minimizes

$$S = [a_1 + a_2 x_0 - y_0]^2 + [a_1 + a_2 x_1 - y_1]^2 + [a_1 + a_2 x_2 - y_2]^2. \tag{1.41}$$

Proof. The conditions that the solution of (1.28) minimize (1.41) are that

$$\frac{\partial^2 S}{\partial a_1^2}, \quad \begin{vmatrix} \dfrac{\partial^2 S}{\partial a_1^2} & \dfrac{\partial^2 S}{\partial a_1\, \partial a_2} \\[2ex] \dfrac{\partial^2 S}{\partial a_1\, \partial a_2} & \dfrac{\partial^2 S}{\partial a_2^2} \end{vmatrix}$$

are both positive. But, from (1.41),

$$\frac{\partial^2 S}{\partial a_1^2} = 6 > 0,$$

while

$$\begin{vmatrix} \dfrac{\partial^2 S}{\partial a_1^2} & \dfrac{\partial^2 S}{\partial a_1\, \partial a_2} \\[2ex] \dfrac{\partial^2 S}{\partial a_1\, \partial a_2} & \dfrac{\partial^2 S}{\partial a_2^2} \end{vmatrix} = \frac{\partial^2 S}{\partial a_1^2}\frac{\partial^2 S}{\partial a_2^2} - \left(\frac{\partial^2 S}{\partial a_1\, \partial a_2}\right)^2$$

$$= 6[2x_0^2 + 2x_1^2 + 2x_2^2] - [2x_0 + 2x_1 + 2x_2]^2$$

$$= 4[(x_0 - x_1)^2 + (x_0 - x_2)^2 + (x_1 - x_2)^2] > 0,$$

so the theorem is proved.

1.8. Remarks

A deeper study of the theory of interpolation can be found in the works of P. J. Davis, Lorentz, and Rice, while various other important topics like Chebyshev, inverse and spline interpolation can be found in P. J. Davis, Goodwin, Greville, Hildebrand (2), Kantorovich and Krylov, Lanczos, Lorentz, Milne (1), Ostrowski, Ralston, Rice, and Todd. It is worth noting that the sharp corner of the graph at $x = 2$ in Figure 1.4 represents a discontinuity of the derivative of the interpolating function on $0 \le x \le 4$, and that such discontinuities, when unpleasant, can be eliminated easily by means of spline interpolation. For an elementary discussion of the special computational difficulties that arise in least-squares interpolation when m, in (1.26), is relatively large, see Ralston.

EXERCISES

1. Let [0,2] be divided into 10 equal parts by the R_{11} set $x_i = 2i/10$, $i = 0, 1, \ldots, 10$. Draw the graph of the discrete function $y_i = f(x_i)$, $x_i \in R_{11}$, for each of the following:

(a) $f(x_i) = x_i$ (b) $f(x_i) = 1 - 3x_i$
(c) $f(x_i) = x_i^2 - 2x_i - 1$ (d) $f(x_i) = \sin \pi x_i$

2. Let [0,1] be divided into 100 equal parts by the R_{101} set $x_i = i/100$, $i = 0, 1, \ldots, 100$. Draw the graph of the discrete function $y_i = f(x_i)$, $x_i \in R_{101}$, for each of the following:
(a) $f(x_i) = 1 - x_i$ (b) $f(x_i) = x_i^2 - 1$ (c) $f(x_i) = 1 - \cos \pi x_i$
3. Find the piecewise linear interpolating function for each of the following discrete functions:
(a) $y_i = x_i^3$ on the R_5 set $x_i = 2i$, $i = 0, 1, 2, 3, 4$
(b) $y_i = x_i^3$ on the R_7 set $x_i = i/10$, $i = 0, 1, 2, 3, 4, 5, 6$
(c) $f(x_i)$ is defined on the R_5 set $x_i = i$, $i = 0, 1, 2, 3, 4$, by $f(x_0) = 8$, $f(x_1) = -3$, $f(x_2) = 2$, $f(x_3) = 0$, $f(x_4) = 0$.
4. Find the piecewise parabolic interpolating function for each discrete function given in Exercise 3.
5. On the R_7 set $x_i = i$, $i = 0, 1, 2, 3, 4, 5, 6$, a discrete function is defined by $f(0) = 0$, $f(1) = 1$, $f(2) = 2$, $f(3) = 4$, $f(4) = 3$, $f(5) = 0$, $f(6) = 1$. Determine (a) the piecewise parabolic interpolating function for the given discrete function, (b) the piecewise cubic interpolating function for the given discrete function (Lagrange form acceptable), and (c) the unique interpolating polynomial of degree six for the given discrete function (Lagrange form acceptable).
6. Given the four points (0,1), (0.1,2), (0.2,0), (0.3,–1), determine (a) the unique cubic polynomial whose graph contains the given points (Lagrange form acceptable), and (b) five polynomials of degree four whose graphs contain the given points. Is there a polynomial of degree two whose graph contains the given points? Why?
7. Estimate the error at $x = \pi/3$ in approximating the function

$$y = \sin x, \qquad 0 \le x \le \pi/2$$

by the parabolic interpolating function through (0,0), $(\pi/4, \sqrt{2}/2)$, $(\pi/2,1)$.
8. Estimate the maximum error in approximating the function

$$y = x + \sin x, \qquad 0 \le x \le 2\pi$$

by the linear interpolating function through (0,0) and $(2\pi,2\pi)$.
9. Estimate the error at $x = \frac{3}{4}$ in approximating the function

$$y = x^3, \qquad 0 \le x \le 1$$

by the parabolic interpolating function through (0,0), $(\frac{1}{2},\frac{1}{8})$, (1,1).
10. By least squares, fit the points (0,–1), (1,1), (2,3), (3,4), (4,5.1) with a straight line $y = a_1 + a_2 x$.

11. By least squares, fit the points

x	-3	-2	-1	0	1	2	3
y	-0.72	-0.01	0.50	0.82	0.89	0.81	0.50

with a polynomial of the form $y = a_1 + a_2x + a_3x^2$.

12. By least squares, fit the points (0,1), (1,0), (2,−7), (3,−26) with a polynomial of the form $y = a_1 + a_2x + a_3x^2 + a_4x^3$ and show that the result is identical with the cubic polynomial through the given points.

13. A discrete function of period 2π is given on $[0,2\pi]$ by

x	0	$\frac{\pi}{4}$	$\frac{\pi}{2}$	$\frac{3\pi}{4}$	π	$\frac{5\pi}{4}$	$\frac{3\pi}{2}$	$\frac{7\pi}{4}$	2π
y	-6	2	5	-1	-2	1	3	-5	-6

By least squares, find a fit of the form

$$y = a_1 \sin x + a_2 \cos x + a_3 \sin 2x + a_4 \cos 2x.$$

14. By least squares, fit the function $y = a_1e^x + a_2e^{2x}$ to the data

x	0	1	2	3
y	-1.1	-7.5	-22.0	-60.1

15. If α and β are constants, prove that

(a) $\sum_{i=1}^{m} (\alpha a_i + \beta b_i) \equiv \alpha \sum_{i=1}^{m} a_i + \beta \sum_{i=1}^{m} b_i$

(b) $\sum_{i=1}^{m} \sum_{j=1}^{n} (a_ib_j) \equiv \sum_{j=1}^{n} \sum_{i=1}^{m} (a_ib_j) \equiv \left(\sum_{i=1}^{m} a_i\right)\left(\sum_{j=1}^{n} b_j\right)$

16. Under the assumptions of Theorem 1.3, what conclusions can be drawn about $a_1, a_2, \ldots, a_{m+1}$ and S for the case $m \geq n + 1$?

17. Let x_0, x_1, x_2, x_3 be an R_4 set. Let $G(x) = a_1 + a_2x + a_3x^2$ be the least-squares fit of four given points (x_0,y_0), (x_1,y_1), (x_2,y_2), (x_3,y_3). Show that this choice of G minimizes

$$S = [a_1 + a_2x_0 + a_3x_0^2 - y_0]^2 + [a_1 + a_2x_1 + a_3x_1^2 - y_1]^2 + [a_1 + a_2x_2 + a_3x_2^2 - y_2]^2 + [a_1 + a_2x_3 + a_3x_3^2 - y_3]^2.$$

18. Chebyshev-type approximation utilizes special polynomials $T_n(x)$ of degree n in x which are defined by

$$T_n(x) = \cos(n \operatorname{arc\,cos} x), \qquad n = 0, 1, 2, \ldots.$$

Show that

$$T_{n+1}(x) = 2xT_n(x) - T_{n-1}(x), \qquad n = 1, 2, \ldots,$$

and then construct T_0, T_1, T_2, T_3, T_4.

19. Extend all concepts and theorems related to discrete functions to the general case in which $[a,b]$ is subdivided into n parts that need not all be of the same length.

Chapter 2

APPROXIMATE INTEGRATION AND DIFFERENTIATION

2.1. Introduction

In physics a large variety of basic quantities, such as work and area, are defined by means of definite integrals of the form

$$\int_a^b f(x)\,dx, \tag{2.1}$$

where $f(x)$ is continuous on $a \leq x \leq b$. The usual method prescribed for evaluating (2.1) is to first find a function $F(x)$ such that $F'(x) \equiv f(x), a \leq x \leq b$, and then to apply the formula

$$\int_a^b f(x)\,dx = F(b) - F(a). \tag{2.2}$$

The function $F(x)$ is called an antiderivative of $f(x)$. Unfortunately it is well known that given an arbitrary $f(x)$, the probability is *zero* that one can construct $F(x)$, so in practice (2.2) becomes virtually useless. For this reason we turn next to numerical methods for approximating (2.1).

Generally speaking, our approach will be first to approximate $f(x)$ on $a \leq x \leq b$ by a function $g(x)$ which can be integrated easily and then to use the approximation

$$\int_a^b g(x)\,dx \cong \int_a^b f(x)\,dx, \tag{2.3}$$

where the symbol $\cong$ denotes the use of an approximation formula. The theory and methods of Chapter 1 will be used extensively in the construction of $g(x)$.

2.2. Trapezoidal Rule

Given $f(x)$ continuous on $a \le x \le b$, subdivide $[a,b]$ into n equal parts by the R_{n+1} set $a = x_0 < x_1 < x_2 < \cdots < x_n = b$ and let $y_i = f(x_i)$. Let $g(x)$ be the piecewise linear interpolating function $L(x)$ of the discrete function $y_i = f(x_i)$, $i = 0, 1, \ldots, n$. Then (2.3) is called the *trapezoidal rule*.

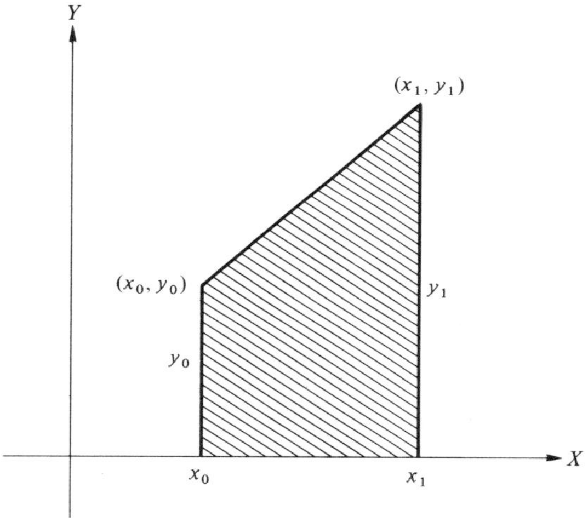

Figure 2.1

Precise formulas for the trapezoidal rule can be developed as follows. The equation of the straight line determined by (x_0,y_0) and (x_1,y_1) in Figure 2.1 is

$$y = y_0 + \frac{y_1 - y_0}{x_1 - x_0}(x - x_0). \tag{2.4}$$

Using (2.4) for $g(x)$ yields

$$\int_{x_0}^{x_1} g(x)\,dx = \left(y_0 x - \frac{y_1 - y_0}{x_1 - x_0} x_0 x + \frac{y_1 - y_0}{x_1 - x_0}\frac{x^2}{2}\right)\Bigg|_{x_0}^{x_1} = \frac{\Delta x}{2}(y_0 + y_1). \tag{2.5}$$

If, as shown in Figure 2.1, both y_0 and y_1 are positive, then (2.5) is the area of the trapezoid shaded in the figure.

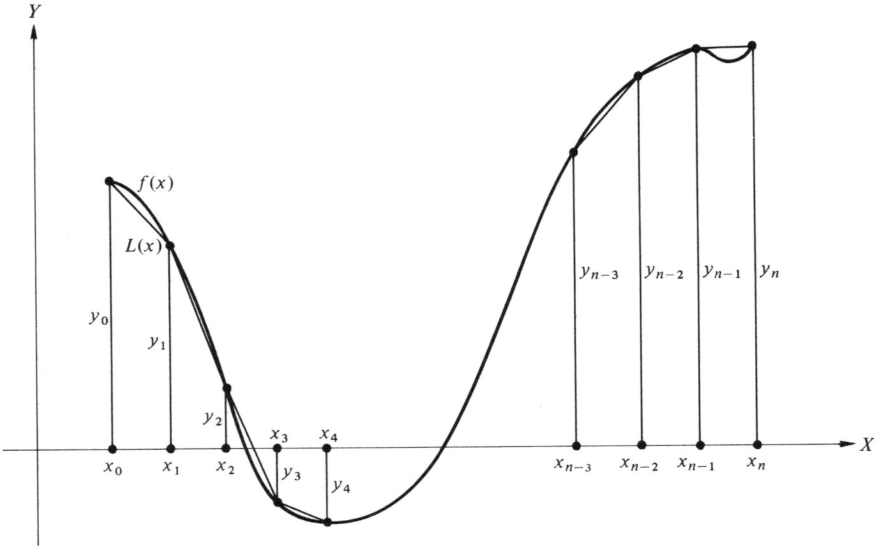

Figure 2.2

In general, with $f(x)$ and $L(x)$ as shown in Figure 2.2, it follows from (2.5) that

$$\int_a^b g(x)\,dx = \int_a^b L(x)\,dx = \sum_{i=0}^{n-1} \int_{x_i}^{x_{i+1}} L(x)\,dx$$

$$= \sum_{i=0}^{n-1} \frac{\Delta x}{2}(y_i + y_{i+1})$$

$$= \frac{\Delta x}{2} \sum_{i=0}^{n-1} [f(x_i) + f(x_{i+1})].$$

Thus the trapezoidal rule takes the form

$$\int_a^b f(x)\,dx \cong \frac{\Delta x}{2} \sum_{i=0}^{n-1} [f(x_i) + f(x_{i+1})]. \tag{2.6}$$

Still another convenient form of (2.6) can be obtained as follows:

$$\sum_{i=0}^{n-1} [f(x_i) + f(x_{i+1})] = [f(x_0) + f(x_1)] + [f(x_1) + f(x_2)]$$
$$+ [f(x_2) + f(x_3)] + \cdots + [f(x_{n-2}) + f(x_{n-1})]$$
$$+ [f(x_{n-1}) + f(x_n)]$$
$$= f(x_0) + 2f(x_1) + 2f(x_2) + \cdots + 2f(x_{n-1}) + f(x_n),$$

so (2.6) can be rewritten as

$$\int_a^b f(x)\,dx \cong \frac{\Delta x}{2}[f(x_0) + 2\sum_{i=1}^{n-1} f(x_i) + f(x_n)]. \tag{2.7}$$

Example 1. Approximate $\int_0^{\pi/2} \sin x\,dx$ by the trapezoidal rule.

Solution. Divide $[0,\pi/2]$ into four equal parts by the R_5 set $x_0 = 0$, $x_1 = \pi/8$, $x_2 = \pi/4$, $x_3 = 3\pi/8$, $x_4 = \pi/2$. Then $y_0 = \sin 0 = 0$, $y_1 = \sin \pi/8 = 0.38268$, $y_2 = \sin \pi/4 = 0.70711$, $y_3 = \sin 3\pi/8 = 0.92388$, $y_4 = \sin \pi/2 = 1.00000$. Thus, by the trapezoidal rule, one obtains the approximation

$$\int_0^{\pi/2} \sin x\,dx \cong \frac{\pi}{16}[0 + 2(0.38268) + 2(0.70711) + 2(0.92388) + 1] = 0.98712. \tag{2.8}$$

The exact value of the given integral is, of course, unity. Division of $[0,\pi/2]$ into 10,000 equal parts and calculating on the CDC 3600 quickly yielded the approximation 0.9999999999.

Example 2. Approximate $\int_0^3 e^{-x^2}\,dx$ by the trapezoidal rule.

Solution. Divide [0,3] into three equal parts by the R_4 set $x_0 = 0$, $x_1 = 1$, $x_2 = 2$, $x_3 = 3$. Then $y_0 = 1$, $y_1 = e^{-1}$, $y_2 = e^{-4}$, $y_3 = e^{-9}$, which yields the approximation

$$\int_0^3 e^{-x^2}\,dx \cong \tfrac{1}{2}[1 + 2(0.367879) + 2(0.018316) + 0.000123] = 0.886257.$$

2.3. Theory for the Trapezoidal Rule

That the trapezoidal rule is a reasonable numerical procedure is a consequence of the theorems that follow.

Theorem 2.1. If $f(x)$ is continuous on $a \le x \le b$, then

$$\lim_{\Delta x \to 0} \left\{ \frac{\Delta x}{2} \sum_{i=0}^{n-1} [f(x_i) + f(x_{i+1})] \right\} = \int_a^b f(x)\,dx. \tag{2.9}$$

Proof. Consider a typical subinterval $[x_i,x_{i+1}]$. If $f(x_i) \le f(x_{i+1})$, then

$$f(x_i) \le \frac{f(x_i) + f(x_{i+1})}{2} \le f(x_{i+1}), \tag{2.10}$$

whereas if $f(x_{i+1}) < f(x_i)$, then

$$f(x_{i+1}) \le \frac{f(x_i) + f(x_{i+1})}{2} \le f(x_i). \tag{2.11}$$

But since f is continuous on $[x_i, x_{i+1}]$, it follows from (2.10) and (2.11) that, in any case, there exists ξ_i such that

$$f(\xi_i) = \frac{f(x_i) + f(x_{i+1})}{2}, \qquad x_i \le \xi_i \le x_{i+1}, \tag{2.12}$$

since f must assume all values between $f(x_i)$ and $f(x_{i+1})$ on the given interval. Hence

$$\lim_{\Delta x \to 0} \left\{ \frac{\Delta x}{2} \sum_{i=0}^{n-1} [f(x_i) + f(x_{i+1})] \right\} = \lim_{\Delta x \to 0} \sum_{i=0}^{n-1} \left[\Delta x \frac{f(x_i) + f(x_{i+1})}{2} \right]$$
$$= \lim_{\Delta x \to 0} \sum_{i=0}^{n-1} [f(\xi_i)\, \Delta x] = \int_a^b f(x)\, dx,$$

and the theorem is proved.

Theorem 2.1 provides the assurance that, as Δx converges to zero, the approximation of the integral by the trapezoidal rule converges to the exact value of the integral. It does *not* tell us, among other things, which of two different approximations corresponding to two different Δx's is the better, nor does it tell us how small we must take Δx to be assured of a given accuracy. To answer some of these questions, we next impose some additional structure on $f(x)$ and derive some interesting consequences.

Suppose $f(x)$ is not only continuous but also has continuous, and hence bounded, first- and second-order derivatives on $a \le x \le b$. Also, let us define the error resulting from application of the trapezoidal rule on subinterval $x_i \le x \le x_{i+1}$ by E_i; that is,

$$E_i = \int_{x_i}^{x_{i+1}} f(x)\, dx - \frac{\Delta x}{2}[f(x_i) + f(x_{i+1})]. \tag{2.13}$$

For convenience, let $\Delta x = h$ and let F be an antiderivative of f so that (2.13) can be rewritten in the form

$$E_i(h) = F(x_i + h) - F(x_i) - \frac{h}{2}[f(x_i + h) + f(x_i)]. \tag{2.13a}$$

Consider now a new function $G(t)$ defined by

$$G(t) = \left\{ F(x_i + t) - F(x_i) - \frac{t}{2}[f(x_i + t) + f(x_i)] \right\} - \frac{t^3}{h^3} E_i(h), \qquad 0 \le t \le h. \tag{2.14}$$

Then

$$G(0) = G(h) = 0.$$

By Rolle's theorem there exists a point α in $0 < \alpha < h$ such that $G'(\alpha) = 0$. But

$$G'(t) = \frac{1}{2}f(x_i + t) - \frac{1}{2}f(x_i) - \frac{t}{2}\frac{df(x_i + t)}{dt} - \frac{3t^2}{h^3}E_i(h),$$

from which it is noted that $G'(0) = 0$ also. Thus

$$G'(0) = G'(\alpha) = 0, \qquad 0 < \alpha < h.$$

Applying Rolle's theorem again, one has that there exists a μ in the range $0 < \mu < \alpha$ at which $G''(\mu) = 0$, or, equivalently, that

$$-\frac{\mu}{2}\frac{d^2f(x_i + \mu)}{dt^2} - \frac{6\mu}{h^3}E_i(h) = 0.$$

Solving this equation for $E_i(h)$ yields

$$E_i(h) = -\frac{h^3}{12}\frac{d^2f(x_i + \mu)}{dt^2}.$$

But, for $x = x_i + t$,

$$\frac{d}{dt}f(x_i + t) = \frac{df(x)}{dx} = f'(x),$$

$$\frac{d^2}{dt^2}f(x_i + t) = \frac{d^2f(x)}{dx^2} = f''(x),$$

so that, finally,

(2.15) $$E_i(h) = -\frac{h^3}{12}f''(\xi_i), \qquad x_i < \xi_i < x_{i+1}.$$

If one defines the error E by

(2.16) $$E = \sum_{i=0}^{n-1} E_i(h),$$

then the following theorem is valid.

Theorem 2.2. In applying the trapezoidal rule to approximate $\int_a^b f(x)\,dx$, where f has continuous second-order derivatives on $[a,b]$, the error E

satisfies the equality

$$E = \frac{(\Delta x)^3}{12} \sum_{i=0}^{n-1} f''(\xi_i), \qquad x_i < \xi_i < x_{i+1}. \tag{2.17}$$

Proof. The proof is a direct consequence of (2.15) and (2.16).

In practice, one can rarely determine E from (2.17), because one rarely knows the ξ_i's. However, (2.17) can often be used to obtain a bound on $|E|$ that can then be used for a variety of practical purposes.

Example 1. Find a bound for $|E|$ in the approximation (2.8).

Solution. It follows from (2.17) that

$$|E| = \frac{1}{12}\left(\frac{\pi}{8}\right)^3 \left| \sum_{i=0}^{3} (-\sin \xi_i) \right|,$$

$$0 < \xi_0 < \frac{\pi}{8} < \xi_1 < \frac{\pi}{4} < \xi_2 < \frac{3\pi}{8} < \xi_3 < \frac{\pi}{2}$$

$$< \frac{1}{12}\left(\frac{\pi}{8}\right)^3 \left(\sin\frac{\pi}{8} + \sin\frac{\pi}{4} + \sin\frac{3\pi}{8} + 1\right)$$

$$< \frac{\pi^3}{6144}(0.39 + 0.71 + 0.93 + 1.00)$$

$$= \frac{3.03\pi^3}{6144} < 0.01616.$$

Example 2. Into how many subdivisions can one divide $0 \le x \le \pi/2$ so that application of the trapezoidal rule to $\int_0^{\pi/2} \sin x\, dx$ will yield a result that is correct to 10^{-3}?

Solution. Let one such subdivision consist of n equal parts. Then, since $|\sin x| < 1$,

$$|E| = \frac{(\Delta x)^3}{12}\left| \sum_{i=0}^{n-1} (-\sin \xi_i) \right| \le \frac{(\Delta x)^3}{12} \sum_{i=0}^{n-1} |\sin \xi_i| \le \frac{n(\Delta x)^3}{12}.$$

But $n\,\Delta x = \pi/2$, so that one wishes

$$\frac{n(\Delta x)^3}{12} = \frac{1}{12}\,\frac{\pi}{2}\left(\frac{\pi}{2n}\right)^2 < 10^{-3},$$

which implies that

$$n > \left(\frac{10^3\pi^3}{96}\right)^{1/2}.$$

Thus any choice $n \ge 18$ would certainly suffice.

2.4. Simpson's Rule

Given $f(x)$ continuous on $a \leq x \leq b$, subdivide $[a,b]$ into n equal parts, where n is *even*, by the R_{n+1} set $a = x_0 < x_1 < x_2 < \cdots < x_n = b$ and let $y_i = f(x_i)$, $i = 0, 1, \ldots, n$. Let $g(x)$ be the piecewise parabolic interpolating function $P(x)$ of discrete function $y_i = f(x_i)$, $i = 0, 1, \ldots, n$. Then (2.3) is called *Simpson's rule*.

Precise formulas for Simpson's rule can be developed as follows. Consider first a particular arrangement of the three points x_0, x_1, x_2 such that $x_0 = -\Delta x$, $x_1 = 0$, $x_2 = \Delta x$, as shown in Figure 2.3. Then, from the Lagrange form, the parabolic function through (x_0,y_0), (x_1,y_1), (x_2,y_2) is

$$y = \frac{1}{2(\Delta x)^2}\{y_0(x^2 - x\,\Delta x) - 2y_1[x^2 - (\Delta x)^2] + y_2(x^2 + x\,\Delta x)\}.$$

Hence

$$\int_{x_0}^{x_2} g(x)\,dx = \int_{x_0}^{x_2} y\,dx = \frac{\Delta x}{3}(y_0 + 4y_1 + y_2). \tag{2.18}$$

But the right-hand side of (2.18) is independent of x_0, x_1, x_2, so no loss of generality resulted from the particular choice in Figure 2.3.

Extending the application of (2.18) to the entire interval $[a,b]$ implies that

$$\int_{x_0}^{x_n} f(x)\,dx \cong \frac{\Delta x}{3}[(y_0 + 4y_1 + y_2) + (y_2 + 4y_3 + y_4) + \cdots + (y_{n-2} + 4y_{n-1} + y_n)], \tag{2.19}$$

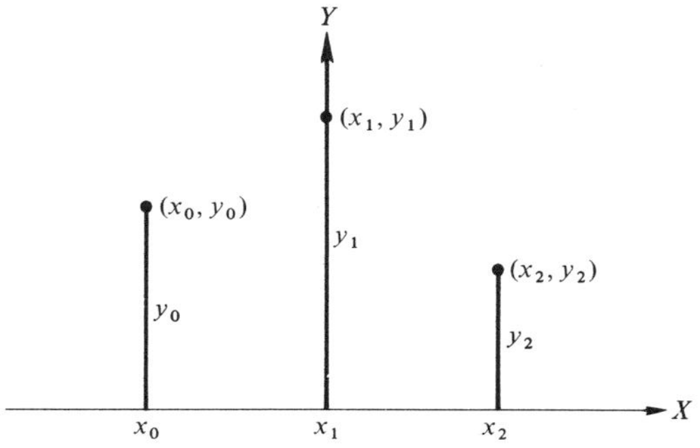

Figure 2.3

This is Simpson's rule, two alternative forms of which are

$$(2.20)\quad \int_{x_0}^{x_n} f(x)\,dx \cong \frac{\Delta x}{3}(y_0 + 4y_1 + 2y_2 + 4y_3 + 2y_4 + 4y_5 + 2y_6 + \cdots + 2y_{n-2} + 4y_{n-1} + y_n)$$

and

$$(2.21)\quad \int_{x_0}^{x_n} f(x)\,dx \cong \frac{\Delta x}{3}\sum_{i=0}^{(n/2)-1}(y_{2i} + 4y_{2i+1} + y_{2i+2}).$$

Of course, each of (2.19) through (2.21) can be rewritten with $f(x_i)$ in place of y_i.

Example. Approximate $\int_0^{\pi/2} \sin x\,dx$ by Simpson's rule.

Solution. Divide $[0, \pi/2]$ into four equal parts by the R_5 set $x_0 = 0$, $x_1 = \pi/8$, $x_2 = \pi/4$, $x_3 = 3\pi/8$, $x_4 = \pi/2$. Then $y_0 = 0$, $y_1 = 0.38268$, $y_2 = 0.70711$, $y_3 = 0.92388$, $y_4 = 1.00000$. By Simpson's rule, then,

$$(2.22)\quad \int_0^{\pi/2} \sin x\,dx \cong \frac{\pi}{24}[0 + 4(0.38268) + 2(0.70711) + 4(0.92388) + 1] = 1.0001.$$

2.5. Theory for Simpson's Rule

That Simpson's rule is a reasonable numerical procedure is a consequence of the following theorems.

Theorem 2.3. If $f(x)$ is continuous on $a \le x \le b$, then

$$\lim_{\Delta x\to 0}\left\{\frac{\Delta x}{3}\sum_{i=0}^{(n/2)-1}[f(x_{2i}) + 4f(x_{2i+1}) + f(x_{2i+2})]\right\} = \int_a^b f(x)\,dx.$$

Proof. Since f is continuous on $[x_{2i}, x_{2i+2}]$, there exists ξ_i in this interval for which

$$6f(\xi_i) = f(x_{2i}) + 4f(x_{2i+1}) + f(x_{2i+2}).$$

Hence

$$\lim_{\Delta x\to 0}\left\{\frac{\Delta x}{3}\sum_{i=0}^{(n/2)-1}[f(x_{2i}) + 4f(x_{2i+1}) + f(x_{2i+2})]\right\}$$

$$= \lim_{\Delta x\to 0}\left\{\frac{\Delta x}{3}\sum_{i=0}^{(n/2)-1}[6f(\xi_i)]\right\}$$

$$= \lim_{\Delta x\to 0}\left\{\sum_{i=0}^{(n/2)-1}[f(\xi_i)][2\,\Delta x]\right\}$$

$$= \lim_{2\Delta x\to 0}\left\{\sum_{i=0}^{(n/2)-1}[f(\xi_i)][2\,\Delta x]\right\}.$$

If one now sets $m = n/2$ and $\Delta t = 2\,\Delta x$, then $\Delta t = (b - a)/m$ and

$$\lim_{2\Delta x \to 0} \left\{ \sum_{i=0}^{(n/2)-1} [f(\xi_i)](2\,\Delta x) \right\} = \lim_{\Delta t \to 0} \left\{ \sum_{i=0}^{m-1} [f(\xi_i)\,\Delta t] \right\}$$

$$= \int_a^b f(t)\,dt = \int_a^b f(x)\,dx,$$

and the theorem is proved.

In order to obtain an error formula analogous to (2.17), set

$$E_{2i} = \int_{x_{2i}}^{x_{2i+2}} f(x)\,dx - \frac{\Delta x}{3}[f(x_{2i}) + 4f(x_{2i+1}) + f(x_{2i+2})]. \tag{2.23}$$

Then, for convenience, set $\bar{x} = x_{2i+1}$ and $\Delta x = h$, and let F be an antiderivative of f, so that (2.23) becomes

$$E_{2i}(h) = F(\bar{x} + h) - F(\bar{x} - h) - \frac{h}{3}[f(\bar{x} - h) + 4f(\bar{x}) + f(\bar{x} + h)]. \tag{2.24}$$

Define $G(t)$ by

$$G(t) = E_{2i}(t) - \frac{E(h)t^5}{h^5}. \tag{2.25}$$

Repeating the argument based on Rolle's theorem which was given to yield (2.17), but this time differentiating through $G'''(t)$, yields

$$E_{2i} = -\frac{h^5}{90} f^{(\mathrm{iv})}(\xi_i), \qquad x_{2i} < \xi_i < x_{2i+2}. \tag{2.26}$$

The following theorem then is valid.

Theorem 2.4. In applying Simpson's rule to approximate $\int_a^b f(x)\,dx$, where f has continuous derivatives up to and including order four on $a \le x \le b$, the error E, defined by

$$E = \sum_{i=0}^{(n/2)-1} E_{2i}, \tag{2.27}$$

satisfies the equality

$$E = \frac{(\Delta x)^5}{90} \sum_{i=0}^{(n/2)-1} f^{(\mathrm{iv})}(\xi_i), \qquad x_{2i} < \xi_i < x_{2i+2}. \tag{2.28}$$

Proof. The proof follows directly from (2.26) and (2.27).

2.6. Approximate Integration by Least Squares

If one wishes to determine $\int_a^b f(x)\,dx$, where $f(x)$ is known only approximately and at only a finite number of points, then one can construct a least-squares fit $g(x)$, which is readily integrable and, if one has a high degree of faith in $g(x)$, use the approximation

$$\int_a^b f(x)\,dx \cong \int_a^b g(x)\,dx. \tag{2.29}$$

Example. A force $f(t)$ is applied during $-3 \le t \le 3$ to accelerate a small particle and one wishes to know the work done, that is,

$$w = \int_{-3}^{3} f(t)\,dt. \tag{2.30}$$

But f is only known experimentally from the following measurements:

$$f(-3) = -0.71, \qquad f(1) = 0.88,$$
$$f(-2) = -0.01, \qquad f(2) = 0.81,$$
$$f(-1) = 0.51, \qquad f(3) = 0.49.$$
$$f(0) = 0.82,$$

However, physically it is surmised that on $-3 \le t \le 3$, f has the form

$$a + bt + ct^2. \tag{2.31}$$

The given data are then fit by least squares with the result

$$g(t) = 0.806 + 0.200t - 0.102t^2, \tag{2.32}$$

and (2.30) is approximated by

$$\int_{-3}^{3} (0.806 + 0.200t - 0.102t^2)\,dt = 3.000.$$

The value of the method of this section depends largely upon one's faith in $g(x)$, which, in turn, depends largely upon the physics of the problem being studied.

2.7. Remarks on Approximate Integration

In practice, one must be very careful not to infer too much from error formulas (2.17) and (2.28). Indeed, if f satisfies the conditions of Theorem 2.2, then $\int_a^b f(x)\,dx$ can be approximated to any accuracy by the trapezoidal rule simply by choosing Δx small enough. The same can be said about Simpson's rule if f satisfies the conditions of Theorem 2.4. But if f satisfies the conditions of Theorem 2.4, then it also satisfies those of Theorem 2.2, and then one can conclude that for Simpson's rule the error goes to zero as $(\Delta x)^5$, whereas for the trapezoidal rule the error goes to zero as $(\Delta x)^3$. Using such arguments, it is often inferred that Simpson's rule is, in general, more economical than, and in this sense superior to, the trapezoidal rule, and in the experiences of many numerical analysts, such a conclusion is correct. Results (2.8) and (2.22) are in agreement with such a conclusion. It is then part of the art of contemporary numerical analysis to choose Simpson's rule first when numerical integration is required.

Nevertheless, it is not always true that Simpson's rule is "superior" to the trapezoidal rule. The first reason is founded in the actual structure of digital computers. The process of taking the limit as Δx goes to zero is simply not realizable in the computer, so any inference based on continuous mathematics simply may have no real meaning in the discrete mathematics of the computer. Second, note that *once* Δx *is fixed*, then formulas (2.17) and (2.28) depend on the derivatives of $f(x)$. In an example like $\int_{0.001}^{1} (1/x^2)\,dx$, upper bounds for the fourth-order derivatives of the integrand are much greater than upper bounds for the second-order derivatives, and the products $(\Delta x)^3 \cdot f''$ and $(\Delta x)^5 \cdot f^{(\mathrm{iv})}$ may be of more significance than $(\Delta x)^3$ and $(\Delta x)^5$ themselves. Third, from the applied-science point of view, singularities, in the form of unbounded derivatives, occur very often, and in such cases (2.17) and (2.28) have no value at all. Typical examples occur in the study of fluid motions, where various derivatives become infinite as one nears the boundary of a container in which the fluid is in motion. Finally, it should be noted that because the ξ values in (2.17) and (2.28) are not known exactly, this shortcoming often leads to the determination of error bounds that are *much too large* to be of significance, and much valuable work has been done in the construction of more reasonable error bounds by the methods of functional analysis (see, e.g., Golomb and Weinberger).

With regard to the use of piecewise cubic, of piecewise quartic, and of higher-order interpolation functions, which had certain value in the era of the desk computer, these offer almost no advantage in the era of the high-speed digital computer and we have therefore avoided the resulting long list of approximate integration formulas.

It should be noted also that almost no major effort has been made in the approximation of multiple integrals except by Monte Carlo methods, which are probabilistic in nature and have been applied with some success in the study of nuclear reactions (see, e.g., Davis and Rabinowitz).

2.8. Approximate Differentiation

Numerical differentiation can be developed in the same spirit as was numerical integration; that is, one can first construct and then differentiate interpolating functions and use the results as approximations. Nevertheless, a second approach, because it yields the same results and because it is of exceptional value in the development of analogues of both ordinary and partial differential equations, will be used instead.

Throughout, let $f(x)$ have continuous derivatives up to and including order four on interval $[a,b]$. Let $[a,b]$ be divided into n equal parts by the R_{n+1} set $x_0, x_1, \ldots, x_n$. Let x_i be a typical point in R_{n+1} and let us first develop approximations for $f'(x_i)$. Consider

$$f'(x_i) \equiv \alpha f(x_i) + \beta f(x_{i+1}), \qquad i < n, \tag{2.33}$$

where α and β are parameters to be determined. Inserting into (2.33) a Taylor expansion for $f(x_{i+1})$ about x_i yields

$$f'(x_i) \equiv \alpha f(x_i) + \beta[f(x_i) + (\Delta x)f'(x_i) + \frac{(\Delta x)^2}{2}f''(\xi)], \qquad x_i < \xi < x_{i+1},$$

from which one has

$$f'(x_i) \equiv f(x_i)[\alpha + \beta] + f'(x_i)[\beta\,\Delta x] + \frac{\beta(\Delta x)^2}{2}f''(\xi). \tag{2.34}$$

Setting corresponding coefficients of (2.34) equal yields

$$\alpha + \beta = 0,$$

$$\beta\,\Delta x = 1,$$

from which one has

$$\beta = \frac{1}{\Delta x}, \qquad \alpha = -\frac{1}{\Delta x}, \tag{2.35}$$

and (2.33) implies that

$$f'(x_i) \cong \frac{1}{\Delta x}[f(x_{i+1}) - f(x_i)], \tag{2.36}$$

while (2.34) implies that the error E in approximation (2.36) is

$$E = \frac{\Delta x}{2}f''(\xi), \qquad x_i < \xi < x_{i+1}. \tag{2.37}$$

Because approximation (2.36) utilizes the two points x_i and x_{i+1}, it is called a two-point, forward-difference approximation of $f'(x_i)$.

In a completely analogous fashion, it follows that a two-point backward-difference approximation is

$$f'(x_i) \cong \frac{1}{\Delta x}[f(x_i) - f(x_{i-1})], \tag{2.38}$$

and the error E in (2.38) is

$$E = -\frac{\Delta x}{2} f''(\xi), \qquad x_{i-1} < \xi < x_i. \tag{2.39}$$

The legitimacy for calling (2.36) and (2.38) *approximations* for $f'(x_i)$ lies in the observation that (2.37) and (2.39) imply that $E \to 0$ as $\Delta x \to 0$.

Example 1. Consider $y = x^2$ on $0 \le x \le 1$. Divide [0,1] into 10 equal parts by the R_{11} set $x_i = (0.1)i$, $i = 0, 1, \ldots, 10$. Then $f'(0.5) = 2(0.5) = 1.0$. The two-point forward-difference approximation (2.36) is

$$f'(0.5) \cong \frac{1}{0.1}[(0.6)^2 - (0.5)^2] = \frac{0.36 - 0.25}{0.1} = 1.1,$$

while the two-point backward-difference approximation (2.38) is

$$f'(0.5) \cong \frac{1}{0.1}[(0.5)^2 - (0.4)^2] = \frac{0.09}{0.1} = 0.9.$$

A three-point forward-difference approximation for $f'(x_i)$ can be developed as follows. Set

$$f'(x_i) \equiv \alpha f(x_i) + \beta f(x_{i+1}) + \gamma f(x_{i+2}), \qquad i < n - 1. \tag{2.40}$$

Substitution of Taylor expansions for $f(x_{i+1})$ and $f(x_{i+2})$ into (2.40) and reorganization of terms yields

$$\begin{aligned} f'(x_i) \equiv\ & f(x_i)[\alpha + \beta + \gamma] + f'(x_i)[\beta\,\Delta x + 2\gamma\,\Delta x] \\ & + f''(x_i)\left[\frac{\beta(\Delta x)^2}{2} + 2\gamma(\Delta x)^2\right] \\ & + \beta\frac{(\Delta x)^3}{6} f'''(\xi_1) + \frac{4}{3}\gamma(\Delta x)^3 f'''(\xi_2), \end{aligned} \tag{2.41}$$

where $x_i < \xi_1 < x_{i+1}$ and $x_i < \xi_2 < x_{i+2}$. Setting corresponding coefficients in (2.41) equal implies

$$\alpha + \beta + \gamma = 0,$$

$$\beta + 2\gamma = \frac{1}{\Delta x},$$

$$\beta + 4\gamma = 0,$$

the solution of which is

$$\alpha = -\frac{3}{2\,\Delta x}, \qquad \beta = \frac{2}{\Delta x}, \qquad \gamma = -\frac{1}{2\,\Delta x}.$$

Thus a three-point forward-difference approximation is

$$f'(x_i) \cong \frac{1}{2\,\Delta x}[-3f(x_i) + 4f(x_{i+1}) - f(x_{i+2})], \tag{2.42}$$

and the error E in approximation (2.42) is

$$E = \frac{(\Delta x)^2}{3} f'''(\xi_1) - \frac{2}{3}(\Delta x)^2 f'''(\xi_2). \tag{2.43}$$

In a fashion completely analogous to that above, a three-point backward-difference approximation is

$$f'(x_i) \cong \frac{1}{2\,\Delta x}[3f(x_i) - 4f(x_{i-1}) + f(x_{i-2})]. \tag{2.44}$$

Finally, suppose one wishes to construct a three-point central-difference approximation for $f'(x_i)$, that is, an approximation of the form

$$f'(x_i) \equiv \alpha f(x_i) + \beta f(x_{i+1}) + \gamma f(x_{i-1}), \qquad i < n. \tag{2.45}$$

Substitution of Taylor expansions into (2.45) implies that

$$\begin{aligned} f'(x_i) \equiv{}& \alpha f(x_i) \\ &+ \beta\left[f(x_i) + (\Delta x) f'(x_i) + \frac{(\Delta x)^2}{2} f''(x_i) + \frac{(\Delta x)^3}{3!} f'''(\xi_1)\right] \\ &+ \gamma\left[f(x_i) - (\Delta x) f'(x_i) + \frac{(\Delta x)^2}{2} f''(x_i) - \frac{(\Delta x)^3}{3!} f'''(\xi_2)\right], \end{aligned} \tag{2.46}$$

where $x_i < \xi_1 < x_{i+1}$ and $x_{i-1} < \xi_2 < x_i$. Recombination of terms in (2.46) yields

$$\begin{aligned} f'(x_i) \equiv{}& f(x_i)[\alpha + \beta + \gamma] + f'(x_i)[\beta\,\Delta x - \gamma\,\Delta x] \\ &+ f''(x_i)\left[\frac{\beta(\Delta x)^2}{2} + \frac{\gamma(\Delta x)^2}{2}\right] \\ &+ \beta\frac{(\Delta x)^3}{6} f'''(\xi_1) - \gamma\frac{(\Delta x)^3}{6} f'''(\xi_2). \end{aligned} \tag{2.47}$$

From the system

$$\alpha + \beta + \gamma = 0,$$

$$\beta - \gamma = \frac{1}{\Delta x},$$

$$\beta + \gamma = 0,$$

one finds

$$\alpha = 0, \qquad \beta = \frac{1}{2\,\Delta x}, \qquad \gamma = -\frac{1}{2\,\Delta x},$$

so that the desired three-point central-difference approximation is

$$f'(x_i) \cong \frac{1}{2\,\Delta x}[f(x_{i+1}) - f(x_{i-1})], \tag{2.48}$$

and the error E in the approximation is

$$E = \frac{(\Delta x)^2}{12} f'''(\xi_1) + \frac{(\Delta x)^2}{12} f'''(\xi_2). \tag{2.49}$$

Example 2. Consider $y = x^2$ on $0 \leq x \leq 1$. Divide [0,1] into 10 equal parts by the R_{11} set $x_i = (0.1)i$, $i = 0, 1, \ldots, 10$. Then $f'(0.5) = 2(0.5) = 1.0$. The three-point forward-difference approximation is

$$f'(0.5) \cong \frac{1}{2(0.1)}[-3(0.5)^2 + 4(0.6)^2 - (0.7)^2] = 1.0,$$

the three-point backward-difference approximation is

$$f'(0.5) \cong \frac{1}{2(0.1)}[3(0.5)^2 - 4(0.4)^2 + (0.3)^2] = 1.0,$$

and the three-point central-difference approximation is

$$f'(0.5) \cong \frac{1}{2(0.1)}[(0.6)^2 - (0.4)^2] = 1.0.$$

In this odd example, then, all three formulas give the same answer, and indeed the resulting approximation is the exact value of $f'(0.5)$. However, for $f = x^2$, one has $f''' \equiv 0$ for all x, so that E must be zero for all three approximations, and the results are, indeed, not so surprising after all.

Because (2.48) does not utilize $f(x_i)$, it is also called a two-point central-difference approximation for $f'(x_i)$.

Next let us find approximations for $f''(x_i)$. For a three-point forward-difference approximation one has

$$\begin{aligned} f''(x_i) &\equiv \alpha f(x_i) + \beta f(x_{i+1}) + \gamma f(x_{i+2}) \\ &= f(x_i)[\alpha + \beta + \gamma] + f'(x_i)[\beta\,\Delta x + 2\gamma\,\Delta x] \\ &\quad + f''(x_i)\left[\frac{\beta}{2}(\Delta x)^2 + 2\gamma(\Delta x)^2\right] + \beta\frac{(\Delta x)^3}{6}f'''(\xi_1) + \frac{4}{3}\gamma(\Delta x)^3 f'''(\xi_2), \end{aligned}$$

where $x_i < \xi_1 < x_{i+1}$ and $x_i < \xi_2 < x_{i+2}$. The solution of the system

$$\alpha + \beta + \gamma = 0,$$

$$\beta + 2\gamma = 0,$$

$$\frac{\beta}{2} + 2\gamma = \frac{1}{(\Delta x)^2}$$

is

$$\alpha = \frac{1}{(\Delta x)^2}, \qquad \beta = -\frac{2}{(\Delta x)^2}, \qquad \gamma = \frac{1}{(\Delta x)^2},$$

so that a three-point forward-difference approximation is

$$f''(x_i) \cong \frac{1}{(\Delta x)^2}[f(x_i) - 2f(x_{i+1}) + f(x_{i+2})], \tag{2.50}$$

and the error E in (2.50) is

$$E = -\tfrac{1}{3}(\Delta x)f'''(\xi_1) + \tfrac{4}{3}(\Delta x)f'''(\xi_2). \tag{2.51}$$

Similarly, a three-point backward-difference formula is

$$f''(x_i) \cong \frac{1}{(\Delta x)^2}[f(x_i) - 2f(x_{i-1}) + f(x_{i-2})]. \tag{2.52}$$

But for a three-point central-difference approximation one finds that

$$\begin{aligned} f''(x_i) &= \alpha f(x_i) + \beta f(x_{i+1}) + \gamma f(x_{i-1}) \\ &= f(x_i)[\alpha + \beta + \gamma] + f'(x_i)[\beta\,\Delta x - \gamma\,\Delta x] \\ &\quad + f''(x_i)\left[\frac{\beta(\Delta x)^2}{2} + \frac{\gamma(\Delta x)^2}{2}\right] + f'''(x_i)\left[\frac{\beta(\Delta x)^3}{6} - \frac{\gamma(\Delta x)^3}{6}\right] \\ &\quad + \frac{\beta(\Delta x)^4}{24}f^{(iv)}(\xi_1) + \frac{\gamma(\Delta x)^4}{24}f^{(iv)}(\xi_2), \end{aligned} \tag{2.53}$$

where $x_i < \xi_1 < x_{i+1}$ and $x_{i-1} < \xi_2 < x_i$. Solution of

$$\alpha + \beta + \gamma = 0,$$
$$\beta - \gamma = 0,$$
$$\beta + \gamma = \frac{2}{(\Delta x)^2}$$

yields

$$\alpha = -\frac{2}{(\Delta x)^2}, \qquad \beta = \frac{1}{(\Delta x)^2}, \qquad \gamma = \frac{1}{(\Delta x)^2}.$$

Thus the three-point central-difference approximation is

$$f''(x_i) \cong \frac{1}{(\Delta x)^2}[f(x_{i-1}) - 2f(x_i) + f(x_{i+1})], \tag{2.54}$$

which is very similar in structure to both (2.50) and (2.52), but this time, from (2.53), the error E in the approximation is

$$E = \frac{(\Delta x)^2}{24} f^{(\mathrm{iv})}(\xi_1) + \frac{(\Delta x)^2}{24} f^{(\mathrm{iv})}(\xi_2). \tag{2.55}$$

The fact that (2.55) is a function of $(\Delta x)^2$ while (2.51) is a function of Δx has led many authors to make the partially valid statement that (2.54) is superior to both (2.50) and (2.52).

Example 3. Consider $y = 1/(1 + x)$ on $0 \le x \le 1$. Divide [0, 1] into 10 equal parts by the R_{11} set $x_i = (0.1)i$, i = 0, 1, . . . , 10. Then

$$f''(0.2) = \frac{1}{0.864} \sim 1.16.$$

The three-point forward approximation is

$$f''(0.2) \cong \frac{1}{0.01}\left(\frac{1}{1.2} - \frac{2}{1.3} + \frac{1}{1.4}\right) \sim 0.916,$$

the three-point backward approximation is

$$f''(0.2) \cong \frac{1}{0.01}\left(\frac{1}{1.2} - \frac{2}{1.1} + \frac{1}{1}\right) \sim 1.515,$$

and the three-point central approximation is

$$f''(0.2) \cong \frac{1}{0.01}\left(\frac{1}{1.1} - \frac{2}{1.2} + \frac{1}{1.3}\right) \sim 1.166.$$

Because Newton's second law of dynamics involves no derivatives of order greater than two, and because our interests are in physical applications, we shall not develop formulas for approximating higher-order derivatives, although these can be constructed easily by the method of this section.

EXERCISES

1. Divide [0,1] into six equal parts and approximate each of the following by the trapezoidal rule:

   (a) $\int_0^1 x^3\,dx$ (b) $\int_0^1 \sin \pi x\,dx$ (c) $\int_0^1 \frac{1}{1+x}\,dx$

2. Find both the exact error and, by means of (2.17), a bound for $|E|$ for each result of Exercise 1.
3. Divide [0, 1] into six equal parts and approximate $\int_0^1 e^{-x^2}\,dx$ by the trapezoidal rule. Can the exact error of the result be given? By means of (2.17), find a bound for $|E|$.
4. By the trapezoidal rule, approximate, correct to one decimal place, each of the following integrals:

   (a) $\int_0^1 x^2\,dx$ (b) $\int_0^1 x^4\,dx$ (c) $\int_0^1 x^2e^{-x}\,dx$

5. Divide [0,1] into six equal parts and approximate each integral of Exercise 1 by Simpson's rule.
6. Find both the exact error and, by means of (2.28), a bound for $|E|$ for each result of Exercise 5.
7. Divide [0,1] into six equal parts and approximate $\int_0^1 e^{-x^2}\,dx$ by Simpson's rule. From (2.28), find a bound for $|E|$.
8. By Simpson's rule, approximate, correct to one decimal place, each of the integrals of Exercise 4.
9. A function $f(x)$ which is integrable on [0,5] is known only at 0, 1, 2, 3, 4, 5 by $f(0) = 4.1$, $f(1) = 6.5$, $f(2) = 9.0$, $f(3) = 13.2$, $f(4) = 17.1$, $f(5) = 20.1$. By least squares, fit the known data with a straight line and then approximate $\int_0^5 f(x)\,dx$ by integrating the resulting linear function.
10. By least squares, fit the known data of Exercise 9 with a parabola and approximate $\int_0^5 f(x)\,dx$ by integrating the resulting quadratic function.

11. For each of the following, determine $y'(1)$ exactly. Then approximate $y'(1)$ by two-point forward-difference formula (2.36). Finally, estimate the error in the approximation by using (2.37). Each function is defined for all real x.
(a) $y = 5, \Delta x = 0.1$ (b) $y = x^3, \Delta x = 0.1$
(c) $y = x^2 - x - 2, \Delta x = 0.1$ (d) $y = x^5 - 2x^2, \Delta x = 0.1$
(e) $y = (1 - x)/(1 + x^2), \Delta x = 0.1$ (f) $y = e^x, \Delta x = 0.05$
(g) $y = \sin \pi x, \Delta x = 0.05$ (h) $y = 2^x, \Delta x = 0.001$
12. For each function in Exercise 11, approximate $y'(1)$ by three-point forward-difference formula (2.42) and estimate the error in the approximation by using (2.43).
13. Show that $y'_i = (1/(6\Delta x))(-11y_i + 18y_{i+1} - 9y_{i+2} + 2y_{i+3})$ is a four-point forward-difference approximation and find the error term.
14. Show that $y'_i = (1/(12\Delta x))(-25y_i + 48y_{i+1} - 36y_{i+2} + 16y_{i+3} - 3y_{i+4})$ is a five-point forward-difference approximation and find the error term.
15. Develop a six-point forward-difference approximation for y'_i.
16. For each of the following, determine $y'(1)$ exactly. Then approximate $y'(1)$ by two-point and three-point backward-difference formulas:
(a) $y = -5, \Delta x = 0.1$ (b) $y = x^4, \Delta x = 0.05$
(c) $y = x^2 - 3x - 7, \Delta x = 0.2$ (d) $y = x^7, \Delta x = 0.05$
(e) $y = (1 - x)/(1 + x^2), \Delta x = 0.1$ (f) $y = e^x, \Delta x = 0.1$
(g) $y = \sin \pi x, \Delta x = 0.001$ (h) $y = e^{x^2}, \Delta x = 0.1$
17. Develop four- and five-point backward-difference formulas for y'_i. How do these compare with the four- and five-point forward-difference formulas given in Exercises 13 and 14? What is the simple way to determine any n-point backward-difference formula for y'_i from the n-point forward-difference formula?
18. For cases (a) through (h) in Exercise 16, determine $y''(1)$ exactly. Then approximate $y''(1)$ by three-point forward-difference formula (2.50) and approximate the error by (2.51).
19. For cases (a) through (h) in Exercise 16, approximate $y''(1)$ by three-point backward-difference formula (2.52).
20. For cases (a) through (h) in Exercise 16, approximate $y''(1)$ by three-point central-difference formula (2.54) and approximate the error by (2.55).
21. Does there exist a two-point forward-difference approximation for $y''(x_i)$? Why?
22. Explain why a Taylor expansion through only f''' was used to develop (2.51), whereas an expansion through $f^{(iv)}$ was used to develop (2.54).

Chapter 3

ALGEBRAIC AND TRANSCENDENTAL SYSTEMS

3.1. Introduction

Digital computers are more closely related to the processes of (classical) algebra than to those of analysis. Fortunately, the approximations for derivatives and integrals developed in Chapter 2 enable one to approximate problems in analysis by ones in algebra. For this reason, we shall study in this chapter two exceptionally viable methods, one of which is direct, the other of which is iterative, which can be applied in the fashion to be shown to solve classes of algebraic and transcendental systems.

3.2. Matrices and Linear Systems

If matrices A, x, and r are defined by

$$x = \begin{pmatrix} x_1 \\ x_2 \\ \vdots \\ x_n \end{pmatrix}, \qquad r = \begin{pmatrix} r_1 \\ r_2 \\ \vdots \\ r_n \end{pmatrix}, \qquad A = \begin{pmatrix} a_{11} & a_{12} & \cdots & a_{1n} \\ a_{21} & a_{22} & \cdots & a_{2n} \\ \vdots & \vdots & & \vdots \\ a_{n1} & a_{n2} & \cdots & a_{nn} \end{pmatrix}, \tag{3.1}$$

then the linear algebraic system

$$\begin{aligned} a_{11}x_1 + a_{12}x_2 + \cdots + a_{1n}x_n &= r_1 \\ a_{21}x_1 + a_{22}x_2 + \cdots + a_{2n}x_n &= r_2 \\ &\vdots \\ a_{n1}x_1 + a_{n2}x_2 + \cdots + a_{nn}x_n &= r_n \end{aligned} \tag{3.2}$$

in the n unknowns $x_1, x_2, \ldots, x_n$ can be written compactly as

$$Ax = r. \tag{3.3}$$

If one assumes that A is nonsingular, then, of course, the vector x exists and is unique. Indeed, its components can be given explicitly in terms of determinants by Cramer's rule. But if one wishes to evaluate the determinants and thereby determine the exact values of $x_1, x_2, \ldots, x_n$, Cramer's rule, although reasonable for $n = 2$, 3, and 4, readily becomes intractable for increasing values of n, and other methods must be introduced. In considering alternative methods, then, let us begin by focusing attention on a class of linear systems of importance in physical applications—tridiagonal systems.

3.3. Tridiagonal Linear Systems

In applied problems, system (3.2) often possesses two important properties, called diagonal dominance and tridiagonal form, which are defined as follows.

Definition 3.1. System (3.2) is said to have *diagonally dominant form*, or is said merely to be *diagonally dominant* if and only if

$$|a_{ii}| \geq \sum_{\substack{j=1 \\ j \neq i}}^{n} |a_{ij}|, \qquad i = 1, 2, \ldots, n, \tag{3.4}$$

with strict inequality holding in (3.4) for at least one value of i.

Example 1

$$\begin{aligned} -2x_1 + x_2 \qquad\qquad &= 7, \\ x_1 - 2x_2 + x_3 \qquad &= -1, \\ x_2 - 2x_3 + x_4 &= 0, \\ x_3 - 2x_4 &= 1 \end{aligned}$$

is diagonally dominant.

Example 2

$$\begin{aligned} 4x_1 + x_2 + x_3 + x_4 &= 1, \\ x_1 + 5x_2 + 2x_3 + 2x_4 &= 7, \\ x_1 - x_2 + 7x_3 - 5x_4 &= 1, \\ 3x_1 + 2x_2 + x_3 + 6x_4 &= -1 \end{aligned}$$

is diagonally dominant.

Example 3. The system

$$\begin{aligned} x_1 - 3x_2 + x_3 &= 2, \\ 3x_1 - x_2 - x_3 &= 1, \\ x_1 + x_2 - 2x_3 &= -1 \end{aligned}$$

is not diagonally dominant, but can be reordered as

$$\begin{aligned} 3x_1 - x_2 - x_3 &= 1, \\ x_1 - 3x_2 + x_3 &= 2, \\ x_1 + x_2 - 2x_3 &= -1, \end{aligned}$$

so that the result is diagonally dominant.

Example 4

$$\begin{aligned} x_1 + x_2 + x_3 + x_4 &= 1, \\ x_1 - x_2 + x_3 - x_4 &= -1, \\ x_1 - x_2 + x_3 + x_4 &= 0, \\ x_1 + x_2 - x_3 - x_4 &= -1 \end{aligned}$$

is not diagonally dominant, nor is any reordering of the system diagonally dominant.

Definition 3.2. System (3.2) is said to have a *tridiagonal form*, or to be *tridiagonal*, if all the elements in A are zero except a_{ii}, $a_{j,j+1}$ and $a_{j+1,j}$, where $i = 1, 2, \ldots, n$; $j = 1, 2, \ldots, n - 1$.

Example

$$\begin{aligned} 3x_1 + x_2 \qquad\qquad\qquad &= 7, \\ x_1 - 2x_2 + x_3 \qquad\qquad &= -1, \\ x_2 + 4x_3 - x_4 \qquad &= -3, \\ x_3 - x_4 + x_5 &= 0, \\ x_4 + 2x_5 &= 1 \end{aligned}$$

is tridiagonal.

The term *tridiagonal* in Definition 3.2 is appropriate because, in matrix form (3.3), A has the particular representation

$$A = \begin{pmatrix} a_{11} & a_{12} & 0 & 0 & 0 & 0 & \cdots & 0 & 0 & 0 & 0 \\ a_{21} & a_{22} & a_{23} & 0 & 0 & 0 & \cdots & 0 & 0 & 0 & 0 \\ 0 & a_{32} & a_{33} & a_{34} & 0 & 0 & \cdots & 0 & 0 & 0 & 0 \\ 0 & 0 & a_{43} & a_{44} & a_{45} & 0 & \cdots & 0 & 0 & 0 & 0 \\ \vdots & \vdots & \vdots & \vdots & \vdots & \vdots & & \vdots & \vdots & \vdots & \vdots \\ 0 & 0 & 0 & 0 & 0 & 0 & \cdots & a_{n-2,\,n-3} & a_{n-2,\,n-2} & a_{n-2,\,n-1} & 0 \\ 0 & 0 & 0 & 0 & 0 & 0 & \cdots & 0 & a_{n-1,\,n-2} & a_{n-1,\,n-1} & a_{n-1,\,n} \\ 0 & 0 & 0 & 0 & 0 & 0 & \cdots & 0 & 0 & a_{n,\,n-1} & a_{nn} \end{pmatrix},$$

in which all elements are zero except those on the main diagonal and on the diagonals just above and just below the main diagonal.

With regard to solving (3.2), the following theorem is of fundamental importance.

Theorem 3.1. Let system (3.2) be tridiagonal and diagonally dominant. If, in addition,

$$a_{ii} < 0, \qquad i = 1, 2, \ldots, n, \tag{3.5}$$

$$a_{j,\,j+1} > 0, \qquad j = 1, 2, \ldots, n-1, \tag{3.6}$$

$$a_{j+1,\,j} > 0, \qquad j = 1, 2, \ldots, n-1, \tag{3.7}$$

then (3.2) has a unique solution.

Proof. Note first that

$$-a_{ii} = |a_{ii}|, \qquad i = 1, 2, \ldots, n, \tag{3.8}$$

$$a_{j,\,j+1} = |a_{j,\,j+1}|, \qquad j = 1, 2, \ldots, n-1, \tag{3.9}$$

$$a_{j+1,\,j} = |a_{j+1,\,j}|, \qquad j = 1, 2, \ldots, n-1. \tag{3.10}$$

Also, for convenience, let $|A|$ represent the determinant of matrix A.

Now, consider the homogeneous system

$$Ax = 0. \tag{3.11}$$

We will show the assumption that (3.11) has a *nonzero* solution vector leads to a contradiction, from which one has $|A| \neq 0$, and the theorem then follows immediately. Without loss of generality, assume one of the components

$x_1, x_2, \ldots, x_n$ of solution vector x is *positive*. [If one is negative, then $-x$ is a solution vector of (3.11) which has a positive component, and one can begin with it.] Since the solution vector has only a finite number of components, it must have a positive maximum component, say x_k. Thus

$$x_k > 0, \quad x_k \geq x_i, \qquad i = 1, 2, \ldots, n. \tag{3.12}$$

If, in (3.12), $k = 1$, then, since (3.2) is tridiagonal,

$$a_{11}x_1 + a_{12}x_2 = 0$$

or

$$x_2 = -\frac{a_{11}x_1}{a_{12}}.$$

From (3.5) and (3.6), one has $x_2 > 0$. But from (3.4), $-a_{11}/a_{12} \geq 1$, so $x_2 \geq x_1$. Finally, since x_1 is maximal, one concludes that

$$x_1 = x_2. \tag{3.13}$$

An analogous discussion for the case $k = n$ implies that

$$x_n = x_{n-1}. \tag{3.14}$$

Finally, suppose that k has one of the values $2, 3, \ldots, n-1$. Then from the tridiagonal structure of (3.2) one has

$$a_{k,\,k-1}x_{k-1} + a_{kk}x_k + a_{k,\,k+1}x_{k+1} = 0. \tag{3.15}$$

Since x_k is maximal, both x_{k-1} and x_{k+1} are less than or equal to x_k. If one, say x_{k-1}, is actually smaller than x_k, then

$$x_{k-1} = x_k - \alpha_1, \qquad \alpha_1 > 0,$$

$$x_{k+1} = x_k - \alpha_2, \qquad \alpha_2 \geq 0,$$

which, upon substitution into (3.15), implies that

$$-a_{kk}x_k = a_{k,\,k-1}x_k + a_{k,\,k+1}x_k - \alpha_1 a_{k,\,k-1} - \alpha_2 a_{k,\,k+1}$$

or

$$-a_{kk} = a_{k,\,k-1} + a_{k,\,k+1} - \alpha_1\frac{a_{k,\,k-1}}{x_k} - \alpha_2\frac{a_{k,\,k+1}}{x_k}.$$

Hence

$$-a_{kk} < a_{k,\,k-1} + a_{k,\,k+1}. \tag{3.16}$$

But $-a_{kk} = |a_{kk}|$, and $a_{k,\,k-1} + a_{k,\,k+1} = |a_{k,\,k-1}| + |a_{k,\,k+1}|$, so (3.16) contradicts (3.4). Thus

$$x_k = x_{k-1} = x_{k+1}. \tag{3.17}$$

Under the assumption, then, that (3.11) has a nonzero solution vector, we conclude from (3.13), (3.14), and (3.17) that

$$x_1 = x_2 = \cdots = x_k = \cdots = x_n > 0. \tag{3.18}$$

But if $x_k, x_k, \ldots, x_k$ are the components of a solution vector x of (3.11), then substitution into (3.11) implies readily that

$$-a_{ii} = \sum_{\substack{j=1 \\ j \neq i}}^{n} a_{ij}, \qquad i = 1, 2, \ldots, n. \tag{3.19}$$

However, (3.8) through (3.10) and (3.19) imply that (3.4) is never valid with strict inequality. This contradicts the assumption of diagonal dominance, and the theorem is proved.

Theorem 3.1 is not only important because it will assure that certain linear algebraic systems have unique solutions, but it will motivate us to construct numerical procedures that satisfy the assumptions of the theorem. Thus, whenever possible, we shall try to deal with matrices that are tridiagonal, diagonally dominant, have negative elements on the main diagonal, and have positive elements just above and just below the main diagonal.

3.4. Direct Solution of Tridiagonal Systems

The actual solution of a linear algebraic system that satisfies the conditions of Theorem 3.1 can be accomplished by both direct and iterative methods. Depending on the quality of the computing facilities available, relatively small systems can be solved by direct methods, whereas relatively large systems must usually be solved by iterative methods. In this section we shall develop one such valuable direct method, and in Section 3.5 we shall develop a valuable iterative technique.

For clarity, consider first a diagonally dominant, tridiagonal system of only five equations in five unknowns, as follows:

$$\begin{aligned}
a_{11}x_1 + a_{12}x_2 \phantom{+ a_{23}x_3 + a_{34}x_4 + a_{45}x_5} &= r_1, \\
a_{21}x_1 + a_{22}x_2 + a_{23}x_3 \phantom{+ a_{34}x_4 + a_{45}x_5} &= r_2, \\
a_{32}x_2 + a_{33}x_3 + a_{34}x_4 \phantom{+ a_{45}x_5} &= r_3, \\
a_{43}x_3 + a_{44}x_4 + a_{45}x_5 &= r_4, \\
a_{54}x_4 + a_{55}x_5 &= r_5.
\end{aligned} \tag{3.20}$$

System (3.20) in matrix form (3.3) would have

$$(3.21)\quad x = \begin{pmatrix} x_1 \\ x_2 \\ x_3 \\ x_4 \\ x_5 \end{pmatrix}, \quad r = \begin{pmatrix} r_1 \\ r_2 \\ r_3 \\ r_4 \\ r_5 \end{pmatrix}, \quad A = \begin{pmatrix} a_{11} & a_{12} & 0 & 0 & 0 \\ a_{21} & a_{22} & a_{23} & 0 & 0 \\ 0 & a_{32} & a_{33} & a_{34} & 0 \\ 0 & 0 & a_{43} & a_{44} & a_{45} \\ 0 & 0 & 0 & a_{54} & a_{55} \end{pmatrix}.$$

Consider first the possibility of writing

$$(3.22)\qquad A = LU,$$

where L and U are simpler than A and have the forms

$$(3.23)\qquad L = \begin{pmatrix} \beta_1 & 0 & 0 & 0 & 0 \\ a_{21} & \beta_2 & 0 & 0 & 0 \\ 0 & a_{32} & \beta_3 & 0 & 0 \\ 0 & 0 & a_{43} & \beta_4 & 0 \\ 0 & 0 & 0 & a_{54} & \beta_5 \end{pmatrix},$$

$$(3.24)\qquad U = \begin{pmatrix} 1 & \gamma_1 & 0 & 0 & 0 \\ 0 & 1 & \gamma_2 & 0 & 0 \\ 0 & 0 & 1 & \gamma_3 & 0 \\ 0 & 0 & 0 & 1 & \gamma_4 \\ 0 & 0 & 0 & 0 & 1 \end{pmatrix}.$$

Then

$$(3.25)\quad LU = \begin{pmatrix} \beta_1 & \beta_1\gamma_1 & 0 & 0 & 0 \\ a_{21} & a_{21}\gamma_1 + \beta_2 & \beta_2\gamma_2 & 0 & 0 \\ 0 & a_{32} & a_{32}\gamma_2 + \beta_3 & \beta_3\gamma_3 & 0 \\ 0 & 0 & a_{43} & a_{43}\gamma_3 + \beta_4 & \beta_4\gamma_4 \\ 0 & 0 & 0 & a_{54} & a_{54}\gamma_4 + \beta_5 \end{pmatrix}.$$

Since one wishes to satisfy (3.22), it follows from (3.21) and (3.25) that the β's and γ's are given constructively by

$$
\begin{aligned}
\beta_1 &= a_{11}, \\
\gamma_1 &= \frac{a_{12}}{\beta_1}, \\
\beta_2 &= a_{22} - a_{21}\gamma_1, \\
\gamma_2 &= \frac{a_{23}}{\beta_2}, \\
\beta_3 &= a_{33} - a_{32}\gamma_2, \\
\gamma_3 &= \frac{a_{34}}{\beta_3}, \\
\beta_4 &= a_{44} - a_{43}\gamma_3, \\
\gamma_4 &= \frac{a_{45}}{\beta_4}, \\
\beta_5 &= a_{55} - a_{54}\gamma_4,
\end{aligned}
\tag{3.26}
$$

or, more concisely, by

$$
\begin{aligned}
\beta_1 &= a_{11} & & \\
\gamma_j &= \frac{a_{j,j+1}}{\beta_j}, & & j = 1, 2, 3, 4, \\
\beta_j &= a_{jj} - a_{j,j-1}\gamma_{j-1}, & & j = 2, 3, 4, 5.
\end{aligned}
\tag{3.27}
$$

Example 1. Let

$$
A = \begin{pmatrix}
-2 & 1 & 0 & 0 & 0 \\
1 & -2 & 1 & 0 & 0 \\
0 & 1 & -2 & 1 & 0 \\
0 & 0 & 1 & -2 & 1 \\
0 & 0 & 0 & 1 & -2
\end{pmatrix}.
$$

Then $a_{ii} = -2$, $i = 1, 2, 3, 4, 5$; $a_{j,j+1} = 1$, $j = 1, 2, 3, 4$; $a_{j,j-1} = 1$, $j = 2, 3, 4, 5$. Thus, from (3.26), $\beta_1 = -2$, $\gamma_1 = -\frac{1}{2}$, $\beta_2 = -\frac{3}{2}$, $\gamma_2 = -\frac{2}{3}$, $\beta_3 = -\frac{4}{3}$, $\gamma_3 = -\frac{3}{4}$, $\beta_4 = -\frac{5}{4}$, $\gamma_4 = -\frac{4}{5}$, $\beta_5 = -\frac{6}{5}$, and the decomposition

(3.22) becomes

$$\begin{pmatrix} -2 & 1 & 0 & 0 & 0 \\ 1 & -2 & 1 & 0 & 0 \\ 0 & 1 & -2 & 1 & 0 \\ 0 & 0 & 1 & -2 & 1 \\ 0 & 0 & 0 & 1 & -2 \end{pmatrix}$$

$$= \begin{pmatrix} -2 & 0 & 0 & 0 & 0 \\ 1 & -\frac{3}{2} & 0 & 0 & 0 \\ 0 & 1 & -\frac{4}{3} & 0 & 0 \\ 0 & 0 & 1 & -\frac{5}{4} & 0 \\ 0 & 0 & 0 & 1 & -\frac{6}{5} \end{pmatrix} \begin{pmatrix} 1 & -\frac{1}{2} & 0 & 0 & 0 \\ 0 & 1 & -\frac{2}{3} & 0 & 0 \\ 0 & 0 & 1 & -\frac{3}{4} & 0 \\ 0 & 0 & 0 & 1 & -\frac{4}{5} \\ 0 & 0 & 0 & 0 & 1 \end{pmatrix}.$$

Now, in order to solve $Ax = r$, we consider the equivalent problem of solving $LUx = r$, for indeed this can be done quite easily as follows. First solve

$$Lz = r \tag{3.28}$$

for the vector z. Then solve

$$Ux = z \tag{3.29}$$

for the vector x. Let us examine why the solution of each system (3.28) and (3.29) is so simple. System (3.28) is

$$\beta_1 z_1 = r_1,$$

$$a_{21} z_1 + \beta_2 z_2 = r_2,$$

$$a_{32} z_2 + \beta_3 z_3 = r_3,$$

$$a_{43} z_3 + \beta_4 z_4 = r_4,$$

$$a_{54} z_4 + \beta_5 z_5 = r_5$$

which, in the exact order written, imply

(3.30)
$$
\begin{aligned}
z_1 &= \frac{r_1}{\beta_1}, \\
z_2 &= \frac{r_2 - a_{21}z_1}{\beta_2}, \\
z_3 &= \frac{r_3 - a_{32}z_2}{\beta_3}, \\
z_4 &= \frac{r_4 - a_{43}z_3}{\beta_4}, \\
z_5 &= \frac{r_5 - a_{54}z_4}{\beta_5},
\end{aligned}
$$

thus generating the solution vector z recursively from z_1. More compactly, (3.30) can be written as

(3.31)
$$
z_1 = \frac{r_1}{\beta_1}, \qquad z_k = \frac{r_k - a_{k,k-1}z_{k-1}}{\beta_k}, \qquad k = 2, 3, 4, 5.
$$

But now that the z's are known, consider (3.29) in complete detail, that is,

$$
\begin{aligned}
x_1 + x_2\gamma_1 &= z_1, \\
x_2 + x_3\gamma_2 &= z_2, \\
x_3 + x_4\gamma_3 &= z_3, \\
x_4 + x_5\gamma_4 &= z_4, \\
x_5 &= z_5.
\end{aligned}
$$

It can be solved recursively by proceeding from the last equation and working up to yield

(3.32)
$$
\begin{aligned}
x_5 &= z_5, \\
x_4 &= z_4 - x_5\gamma_4, \\
x_3 &= z_3 - x_4\gamma_3, \\
x_2 &= z_2 - x_3\gamma_2, \\
x_1 &= z_1 - x_2\gamma_1,
\end{aligned}
$$

thus yielding the desired solution vector x. An equivalent form of (3.32) is

(3.33)
$$
x_5 = z_5, \quad x_j = z_j - x_{j+1}\gamma_j, \qquad j = 4, 3, 2, 1.
$$

Example 2. Solve the system

$$
\begin{aligned}
-2x_1 + x_2 \qquad\qquad\qquad &= 1,\\
x_1 - 2x_2 + x_3 \qquad\qquad &= 0,\\
x_2 - 2x_3 + x_4 \qquad &= 0,\\
x_3 - 2x_4 + x_5 &= 0,\\
x_4 - 2x_5 &= 0.
\end{aligned}
$$

Solution. In the previous example, it was shown that

$$
\begin{aligned}
a_{ii} &= -2, \qquad i = 1, 2, 3, 4, 5,\\
a_{j,j+1} &= 1, \qquad j = 1, 2, 3, 4,\\
a_{j,j-1} &= 1, \qquad j = 2, 3, 4, 5,
\end{aligned}
$$

$$\beta_1 = -2, \beta_2 = -\tfrac{3}{2}, \beta_3 = -\tfrac{4}{3}, \beta_4 = -\tfrac{5}{4}, \beta_5 = -\tfrac{6}{5},$$

$$\gamma_1 = -\tfrac{1}{2}, \gamma_2 = -\tfrac{2}{3}, \gamma_3 = -\tfrac{3}{4}, \gamma_4 = -\tfrac{4}{5}.$$

Thus, from (3.30),

$$
\begin{aligned}
z_1 &= -\tfrac{1}{2},\\
z_2 &= -\tfrac{1}{3},\\
z_3 &= -\tfrac{1}{4},\\
z_4 &= -\tfrac{1}{5},\\
z_5 &= -\tfrac{1}{6},
\end{aligned}
$$

and finally, from (3.32),

$$
\begin{aligned}
x_5 &= -\tfrac{1}{6},\\
x_4 &= -\tfrac{1}{3},\\
x_3 &= -\tfrac{1}{2},\\
x_2 &= -\tfrac{2}{3},\\
x_1 &= -\tfrac{5}{6}.
\end{aligned}
$$

The direct method of this section can now be extended easily from (3.27), (3.31), and (3.33) to n equations in n unknowns as follows. Consider a tridiagonal system that satisfies the conditions of Theorem 4.1. Then the direct method for generating solution vector x consists of the following three-step algorithm:

1. Generate $\beta_1, \beta_2, \ldots, \beta_n$ and $\gamma_1, \gamma_2, \ldots, \gamma_{n-1}$ recursively from

$$\beta_1 = a_{11},$$

$$\gamma_j = \frac{a_{j,j+1}}{\beta_j}, \qquad j = 1, 2, \ldots, n-1,$$

$$\beta_j = a_{jj} - a_{j,j-1}\gamma_{j-1}, \qquad j = 2, 3, \ldots, n.$$

2. Generate vector z recursively from

$$z_1 = \frac{r_1}{\beta_1}, \quad z_j = \frac{r_j - a_{j,j-1}z_{j-1}}{\beta_j}, \qquad j = 2, 3, \ldots, n.$$

3. Generate solution vector x recursively from

$$x_n = z_n, \quad x_k = z_k - x_{k+1}\gamma_k, \qquad k = n-1, n-2, \ldots, 3, 2, 1.$$

3.5. Iterative Solution of Linear and Nonlinear Systems

We shall consider next an extraordinarily general *iterative* technique which can be used to solve tridiagonal systems, more general classes of linear systems, and certain important classes of nonlinear systems. The method to be developed is called the *generalized Newton's method* and when it is applied to linear systems it is called *successive overrelaxation* (SOR). When it is applied to linear systems with a unit successive overrelaxation factor, it is called *Gauss–Seidel iteration.* Typical of its utility is that it can be applied to solve linear systems related to potential theory problems which consist of 25,000 equations in as many unknowns in 13 minutes on the UNIVAC 1108 [Greenspan (6)].

Suppose first that one wishes to find a root of the equation

$$f(x) = 0, \tag{3.34}$$

where f is a given differentiable function. Let the graph of

$$y = f(x) \tag{3.35}$$

be as shown in Figure 3.1. The problem of finding a root of (3.34) is, of course, completely equivalent to that of finding a zero of (3.35), that is, of finding a point where the graph of (3.35) intersects the X axis, and we shall try to do the latter as follows.

Suppose that $\bar{x}$, as shown in Figure 3.1, is a zero of $f(x)$. Since $\bar{x}$ is, in general, not known, let us make a guess at it (called an initial guess), say $x = x^{(1)}$. If $f(x^{(1)}) = 0$, then the problem is solved. In all probability, however, $f(x^{(1)}) \neq 0$, and one proceeds as follows to try to get a new value $x^{(2)}$ which

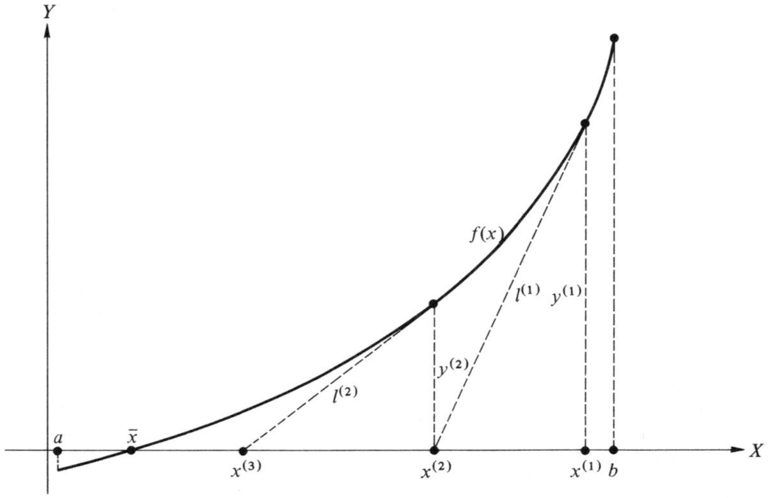

Figure 3.1

approximates $\bar{x}$ better than does $x^{(1)}$. Let $f(x^{(1)}) = y^{(1)}$. Then, at the point $(x^{(1)}, y^{(1)})$, the slope of the tangent line $l^{(1)}$ to $f(x)$ is $f'(x^{(1)})$. Thus the equation of the line $l^{(1)}$ is

$$y - f(x^{(1)}) = f'(x^{(1)}) \cdot (x - x^{(1)}). \tag{3.36}$$

Let $x^{(2)}$ be the point in which $l^{(1)}$ intersects the X axis. Analytically $x^{(2)}$ can be determined from (3.36) by setting $y = 0$, that is, from

$$-f(x^{(1)}) = f'(x^{(1)}) \cdot (x^{(2)} - x^{(1)}). \tag{3.37}$$

Under the assumption that $f'(x^{(1)}) \neq 0$, that is, that $l^{(1)}$ is not parallel to the X axis, it follows from (3.37) that

$$x^{(2)} = x^{(1)} - \frac{f(x^{(1)})}{f'(x^{(1)})}. \tag{3.38}$$

If $f(x^{(2)}) = 0$, then the problem is solved. If $f(x^{(2)}) \neq 0$, then one proceeds to try to improve on the approximation $x^{(2)}$. Let $f(x^{(2)}) = y^{(2)}$. Let $l^{(2)}$ be the tangent line to $f(x)$ at $(x^{(2)}, y^{(2)})$. Let $x^{(3)}$ be the intersection of $l^{(2)}$ with the X axis, so that, as in the derivation of (3.38), one finds

$$x^{(3)} = x^{(2)} - \frac{f(x^{(2)})}{f'(x^{(2)})}, \qquad f'(x^{(2)}) \neq 0. \tag{3.39}$$

Again, if $f(x^{(3)}) = 0$, then the problem is solved. If $f(x^{(3)}) \neq 0$, then proceed to construct $x^{(4)}, x^{(5)}, \ldots$, in the same spirit as $x^{(2)}$ and $x^{(3)}$ were constructed above. After n steps a value $x^{(n+1)}$ is determined by the formula

$$x^{(n+1)} = x^{(n)} - \frac{f(x^{(n)})}{f'(x^{(n)})}, \qquad f'(x^{(n)}) \neq 0. \tag{3.40}$$

The formula (3.40) is called *Newton's formula* and the entire iterative procedure described above is called *Newton's method.*

Newton's method can be used under suitable conditions [Ostrowski; Rall (2)] to approximate a root quite accurately by hand and, indeed, to almost any degree of accuracy with great speed on a computer.

Example 1. Approximate the positive root of the equation

$$x^3 + \sqrt{3}x^2 - 2x - 2\sqrt{3} = 0$$

by Newton's method.

Solution. If one sets

$$f(x) = x^3 + \sqrt{3}x^2 - 2x - 2\sqrt{3},$$

then

$$f(0) = -2\sqrt{3} < 0,$$

$$f(1) = 1 + \sqrt{3} - 2 - 2\sqrt{3} = -1 - \sqrt{3} < 0,$$

$$f(2) = 8 + 4\sqrt{3} - 4 - 2\sqrt{3} = 4 + 2\sqrt{3} > 0,$$

and a root lies between 1 and 2. Newton's formula has the form

$$x^{(n+1)} = x^{(n)} - \frac{[x^{(n)}]^3 + \sqrt{3}[x^{(n)}]^2 - 2x^{(n)} - 2\sqrt{3}}{3[x^{(n)}]^2 + 2\sqrt{3}x^{(n)} - 2}.$$

Approximating $\sqrt{3}$ by 1.7, setting $x^{(1)} = 2.0$, and rounding to one decimal place, one finds

$$x^{(2)} = 1.6,$$

$$x^{(3)} = 1.4,$$

$$x^{(4)} = 1.4.$$

Since $x^{(3)} = x^{(4)}$, the Newton iteration formula will yield no new value of x, and the approximation that results is $x = 1.4$. The exact solution is $x = \sqrt{2}$.

Example 2. Newton's method was programmed for the CDC 3600 to find a positive root of the equation in Example 1. To make it difficult for the computer, the initial guess was taken to be $x^{(1)} = 100{,}000$ and numbers were rounded to nine decimal places. The iteration stopped with $x^{(32)} = x^{(33)} = 1.414213562$. The running time was 4 milliseconds.

For more difficult problems than those considered in the above examples, it will be of value to have a method that can yield a root in fewer iterations than those determined by Newton's method. Such a method can be developed as follows. Suppose, instead of constructing the line $l^{(1)}$ shown in Figure 3.1, one were clever enough to construct a different line through $(x^{(1)}, y^{(1)})$, which would intersect the X axis closer to $\bar{x}$ than is $x^{(2)}$. Such a line would have an equation of the form

$$y - f(x^{(1)}) = \tau \cdot f'(x^{(1)}) \cdot (x - x^{(1)}), \qquad \tau \neq 1, \tag{3.41}$$

for indeed the line would differ from $l^{(1)}$ only in slope. Setting $y = 0$ and $x = x^{(2)}$ in (3.41) would then yield

$$x^{(2)} = x^{(1)} - \frac{1}{\tau}\frac{f(x^{(1)})}{f'(x^{(1)})}, \qquad \tau \cdot f'(x^{(1)}) \neq 0, \tag{3.42}$$

and just as the Newton's formula (3.40) was developed by first considering (3.38), then from (3.42) would follow the formula

$$x^{(n+1)} = x^{(n)} - \frac{1}{\tau}\frac{f(x^{(n)})}{f'(x^{(n)})}, \qquad \tau \cdot f'(x^{(n)}) \neq 0. \tag{3.43}$$

If, for simplicity, one sets $\omega = 1/\tau$, then (3.43) takes the form

$$x^{(n+1)} = x^{(n)} - \omega\frac{f(x^{(n)})}{f'(x^{(n)})}, \qquad f'(x^{(n)}) \neq 0 \tag{3.44}$$

and is called the *generalized Newton's formula*. The constant ω is called an overrelaxation factor. The iterative method which then uses (3.44) in place of (3.40) is called the *generalized Newton's method*. And, of course, Newton's formula results from (3.44) in the special case $\omega = 1$.

Example 3. Approximate the positive root of

$$x^3 + \sqrt{3}x^2 - 2x - 2\sqrt{3} = 0$$

by the generalized Newton's method.

Solution. The generalized Newton's formula for the given equation is

$$x^{(n+1)} = x^{(n)} - \omega \frac{[x^{(n)}]^3 + \sqrt{3}[x^{(n)}]^2 - 2x^{(n)} - 2\sqrt{3}}{3[x^{(n)}]^2 + 2\sqrt{3}x^{(n)} - 2}. \tag{3.45}$$

Approximating $\sqrt{3}$ by 1.7, setting $x^{(1)} = 2.0$ and $\omega = 1.3$, and rounding to one decimal place, one has from (3.45) that

$$x^{(2)} = 1.4,$$

$$x^{(3)} = 1.4.$$

Since $x^{(2)} = x^{(3)}$, the iteration will yield no change in the approximation $x = 1.4$. Note that this result was obtained with fewer iterations than were required in Example 1, that is, with $\omega = 1$.

Suppose now that we have to solve two equations in two unknowns, say

$$f_1(x_1,x_2) = 0, \tag{3.46}$$

$$f_2(x_1,x_2) = 0. \tag{3.47}$$

Then a straightforward extension of (3.44) which we shall use for the system (3.46) and (3.47) is

$$x_1^{(n+1)} = x_1^{(n)} - \omega \frac{f_1(x_1^{(n)},x_2^{(n)})}{\partial f_1(x_1^{(n)},x_2^{(n)})/\partial x_1}, \tag{3.48}$$

$$x_2^{(n+1)} = x_2^{(n)} - \omega \frac{f_2(x_1^{(n+1)},x_2^{(n)})}{\partial f_2(x_1^{(n+1)},x_2^{(n)})/\partial x_2}. \tag{3.49}$$

Note that formulas (3.48) and (3.49) do not contain matrices, as many other generalizations of Newton's method do, and that the result $x_1^{(n+1)}$, not $x_1^{(n)}$, is used to calculate $x_2^{(n+1)}$.

Example 4. Consider the tridiagonal linear system of equations

$$-2x_1 + x_2 = 3,$$

$$x_1 - 2x_2 = -3.$$

Set

$$f_1(x_1,x_2) = -2x_1 + x_2 - 3,$$

$$f_2(x_1,x_2) = x_1 - 2x_2 + 3.$$

Then (3.48) and (3.49) take the forms

$$x_1^{(n+1)} = x_1^{(n)} - \omega \frac{-2x_1^{(n)} + x_2^{(n)} - 3}{-2}, \tag{3.50}$$

$$x_2^{(n+1)} = x_2^{(n)} - \omega \frac{x_1^{(n+1)} - 2x_2^{(n)} + 3}{-2}. \tag{3.51}$$

With an initial guess of $x_1^{(1)} = x_2^{(1)} = 0$ and with $\omega = 1$, one generates from (3.50) and (3.51) the sequences $x_1^{(n+1)}$ and $x_2^{(n+1)}$ as follows:

$$x_1^{(2)} = -\tfrac{3}{2}, x_2^{(2)} = \tfrac{3}{4}, x_1^{(3)} = -\tfrac{9}{8}, x_2^{(3)} = \tfrac{15}{16}, x_1^{(4)} = -\tfrac{33}{32}, x_2^{(4)} = \tfrac{33}{32}, \ldots.$$

Thus, for $x_1^{(n+1)}$, one has $x_1^{(1)} = 0, x_1^{(2)} = -\frac{3}{2}, x_1^{(3)} = -\frac{9}{8}, x_1^{(4)} = -\frac{33}{32}$, which does seem to be converging to the correct value $x_1 = -1$; for $x_2^{(n+1)}$, one has $x_2^{(1)} = 0$, $x_2^{(2)} = \frac{3}{4}$, $x_2^{(3)} = \frac{15}{16}$, $x_2^{(4)} = \frac{31}{32}$, which does seem to be converging to the correct value $x_2 = 1$.

Example 5. Consider the system of transcendental equations

$$e^{x_1} + x_1 - 3x_2 - 3 = 0,$$

$$e^{x_2} + x_2 - 2x_1 + 1 = 0.$$

Set

$$f_1(x_1,x_2) = e^{x_1} + x_1 - 3x_2 - 3,$$

$$f_2(x_1,x_2) = e^{x_2} + x_2 - 2x_1 + 1.$$

Then the generalized Newton's formulas for the system take the particular form

$$x_1^{(n+1)} = x_1^{(n)} - \omega \frac{e^{x_1^{(n)}} + x_1^{(n)} - 3x_2^{(n)} - 3}{e^{x_1^{(n)}} + 1}, \tag{3.52}$$

$$x_2^{(n+1)} = x_2^{(n)} - \omega \frac{e^{x_2^{(n)}} + x_2^{(n)} - 2x_1^{(n+1)} + 1}{e^{x_2^{(n)}} + 1}. \tag{3.53}$$

If one sets $x_1^{(1)} = x_2^{(1)} = 0$ and $\omega = 1.5$, then (3.52) and (3.53) imply that

$$x_1^{(2)} = x_1^{(1)} - 1.5 \frac{e^{x_1^{(1)}} + x_1^{(1)} - 3x_2^{(1)} - 3}{e^{x_1^{(1)}} + 1} = 1.5, \tag{3.54}$$

$$x_2^{(2)} = x_2^{(1)} - 1.5 \frac{e^{x_2^{(1)}} + x_2^{(1)} - 2x_1^{(2)} + 1}{e^{x_2^{(1)}} + 1} = 0.75. \tag{3.55}$$

The results (3.54) and (3.55) would then be inserted into (3.52) and (3.53) to produce $x_1^{(3)}$ and $x_2^{(3)}$, and the iteration would continue in this fashion.

Finally, let us extend (3.46) and (3.47) to a system of k equations in k unknowns. Suppose one has to solve a system of equations that has the following form:

(3.56) $$f_1(x_1,x_2,x_3,\ldots,x_{k-1},x_k) = 0,$$

(3.57) $$f_2(x_1,x_2,x_3,\ldots,x_{k-1},x_k) = 0,$$

(3.58) $$f_3(x_1,x_2,x_3,\ldots,x_{k-1},x_k) = 0,$$

$$\vdots$$

(3.59) $$f_{k-1}(x_1,x_2,x_3,\ldots,x_{k-1},x_k) = 0,$$

(3.60) $$f_k(x_1,x_2,x_3,\ldots,x_{k-1},x_k) = 0.$$

Then the generalized Newton's formulas for the system (3.56) through (3.60) which will be of particular value to us are

(3.61)

$$x_1^{(n+1)} = x_1^{(n)} - \omega \frac{f_1(x_1^{(n)},x_2^{(n)},x_3^{(n)},\ldots,x_{k-1}^{(n)},x_k^{(n)})}{\partial f_1(x_1^{(n)},x_2^{(n)},x_3^{(n)},\ldots,x_{k-1}^{(n)},x_k^{(n)})/\partial x_1},$$

(3.62)

$$x_2^{(n+1)} = x_2^{(n)} - \omega \frac{f_2(x_1^{(n+1)},x_2^{(n)},x_3^{(n)},\ldots,x_{k-1}^{(n)},x_k^{(n)})}{\partial f_2(x_1^{(n+1)},x_2^{(n)},x_3^{(n)},\ldots,x_{k-1}^{(n)},x_k^{(n)})/\partial x_2},$$

(3.63)

$$x_3^{(n+1)} = x_3^{(n)} - \omega \frac{f_3(x_1^{(n+1)},x_2^{(n+1)},x_3^{(n)},\ldots,x_{k-1}^{(n)},x_k^{(n)})}{\partial f_3(x_1^{(n+1)},x_2^{(n+1)},x_3^{(n)},\ldots,x_{k-1}^{(n)},x_k^{(n)})/\partial x_3},$$

$$\vdots$$

(3.64)

$$x_{k-1}^{(n+1)} = x_{k-1}^{(n)} - \omega \frac{f_{k-1}(x_1^{(n+1)},x_2^{(n+1)},x_3^{(n+1)},\ldots,x_{k-2}^{(n+1)},x_{k-1}^{(n)},x_k^{(n)})}{\partial f_{k-1}(x_1^{(n+1)},x_2^{(n+1)},x_3^{(n+1)},\ldots,x_{k-2}^{(n+1)},x_{k-1}^{(n)},x_k^{(n)})/\partial x_{k-1}},$$

(3.65)

$$x_k^{(n+1)} = x_k^{(n)} - \omega \frac{f_k(x_1^{(n+1)},x_2^{(n+1)},\ldots,x_{k-1}^{(n+1)},x_k^{(n)})}{\partial f_k(x_1^{(n+1)},x_2^{(n+1)},\ldots,x_{k-1}^{(n+1)},x_k^{(n)})/\partial x_k}.$$

The application of (3.61) through (3.65) to system (3.56) through (3.60) will be called the generalized Newton's method for systems.

3.6. Remarks

The number of methods under study currently for the direct or iterative solution of systems of equations is voluminous, the reasons being that no method applies universally and that different methods are preferable for systems with different structural properties. The interested reader would benefit from an examination of the Gauss–Jordan elimination method, matrix inversion, relaxation, the Crout method, the square-root method, the method of postmultiplication, the gradient method, general iteration, line and block overrelaxation, Richardson's method, and the Peaceman–Rachford method (see, e.g., Forsythe, Froberg, Goodwin, Kunz, Ostrowski, and Varga).

From a practical point of view, it should be noted also that direct methods, because of an inherent step-by-step accumulation of roundoff error, are in general restricted in applicability to systems of fewer equations than are iterative methods, so an iterative method is preferred if one must solve a relatively large system of equations.

EXERCISES

1. Which of the following systems are, or can be put, in a form that is diagonally dominant?

(a)
$$\begin{aligned} 2x_1 - 2x_2 \qquad\quad &= 1\\ x_1 - 2x_2 + x_3 &= 1\\ 2x_2 - 2x_3 &= 1 \end{aligned}$$

(b)
$$\begin{aligned} x_1 + x_2 - 2x_3 &= 0\\ x_1 - 2x_2 + x_3 &= 1\\ -2x_1 + x_2 + x_3 &= 0 \end{aligned}$$

(c)
$$\begin{aligned} -4x_1 + x_2 + x_3 + x_4 &= 1\\ x_1 - 5x_2 + x_3 + x_4 &= 0\\ x_1 + x_2 - 6x_3 + x_4 &= 1\\ x_1 + x_2 + x_3 - 7x_4 &= 0 \end{aligned}$$

(d)
$$\begin{aligned} -4x_1 + x_2 + x_3 + x_4 &= 1\\ -5x_2 + x_3 + x_4 &= 0\\ -6x_3 + x_4 &= 1\\ -7x_4 &= 0 \end{aligned}$$

(e)
$$\begin{aligned} -4x_1 + x_2 \qquad\qquad &= 1\\ x_1 - 5x_2 + x_3 \qquad &= 0\\ x_2 - 6x_3 + x_4 &= 1\\ x_3 - 7x_4 &= 0 \end{aligned}$$

(f)
$$\begin{aligned} -4x_1 \qquad &= 1\\ -5x_2 \qquad &= 0\\ -6x_3 \qquad &= 1\\ -7x_4 &= 0 \end{aligned}$$

2. Which of the systems in Exercise 1 are, or can be rewritten, in tridiagonal form?
3. Write each of the following tridiagonal matrices as a product LU, where L is lower triangular and U is upper triangular:

(a) $\begin{pmatrix} -2 & 1 & 0 & 0 \\ 1 & -2 & 1 & 0 \\ 0 & 1 & -2 & 1 \\ 0 & 0 & 1 & -2 \end{pmatrix}$ (b) $\begin{pmatrix} -3 & 1 & 0 & 0 \\ 2 & -3 & 1 & 0 \\ 0 & 2 & -3 & 1 \\ 0 & 0 & 2 & -3 \end{pmatrix}$

(c) $\begin{pmatrix} 10 & -3 & 0 & 0 \\ 4 & 10 & -3 & 0 \\ 0 & 4 & 10 & -3 \\ 0 & 0 & 4 & 10 \end{pmatrix}$ (d) $\begin{pmatrix} 5 & -2 & 0 & 0 & 0 \\ -2 & 5 & -2 & 0 & 0 \\ 0 & -2 & 5 & -2 & 0 \\ 0 & 0 & -2 & 5 & -2 \\ 0 & 0 & 0 & -2 & 5 \end{pmatrix}$

4. For (3.26), show that none of β_1, β_2, β_3, β_4 is zero.
5. Solve each of the following tridiagonal systems by the direct method:

(a)
$$\begin{aligned} -2x_1 + x_2 \qquad\qquad &= -2 \\ x_1 - 2x_2 + x_3 &= 1 \\ x_2 - 2x_3 + x_4 &= 0 \\ x_3 - 2x_4 &= 0 \end{aligned}$$

(b)
$$\begin{aligned} 5x_1 - 2x_2 \qquad\qquad &= 5 \\ -2x_1 + 5x_2 - 2x_3 &= -2 \\ -2x_2 + 5x_3 - 2x_4 &= 0 \\ -2x_3 + 5x_4 - 2x_5 &= 2 \\ -2x_4 + 5x_5 &= -5 \end{aligned}$$

(c)
$$\begin{aligned} -2x_1 + x_2 \qquad\qquad &= 1 \\ x_1 - 4x_2 + x_3 \qquad &= 0 \\ x_2 - 3x_3 + x_4 &= -1 \\ x_3 - 2x_4 &= 0 \end{aligned}$$

(d)
$$\begin{aligned} -2x_1 + x_2 \qquad\qquad &= -2 \\ x_1 - 5x_2 + x_3 \qquad &= 1 \\ x_2 - 3x_3 + x_4 &= 0 \\ x_3 - 2x_4 &= 0 \end{aligned}$$

(e)
$$\begin{aligned} -3x_1 + x_2 \qquad\qquad &= 1 \\ x_1 - 2x_2 + x_3 \qquad &= 0 \\ x_2 - 5x_3 + x_4 &= 1 \\ x_3 - 7x_4 &= 0 \end{aligned}$$

6. Find $x_1^{(6)}$ and $x_2^{(6)}$ by the generalized Newton's method for each of the following systems. Compute to three decimal places and use $x_1^{(1)} = x_2^{(1)} = 0$, $\omega = 1$.

(a) $4x_1 - x_2 = 12$, $x_1 - 3x_2 = 3$ (b) $10x_1 + 5x_2 = 5$, $x_1 - 6x_2 = 7$ (c) $4x_1 - x_2 = e^{x_1}$, $x_1 - 3x_2 = e^{x_2}$

7. Repeat Exercise 6 with $\omega = 1.8$.
8. Find $x_1^{(5)}, x_2^{(5)}, x_3^{(5)}$ by the generalized Newton's method for each of the following systems. Compute to two decimal places and use $x_1^{(1)} = x_2^{(1)} = x_3^{(1)} = 0$, $\omega = 1.0$:

(a)
$$\begin{aligned} 5x_1 - x_2 + x_3 &= 2 \\ x_1 - 4x_2 + x_3 &= 0 \\ 2x_1 + x_2 - 4x_2 &= 6 \end{aligned}$$

(b)
$$\frac{x_1}{\sqrt{1 + (x_1/0.2)^2}} - \frac{(\cos^2 \pi/16)(x_2 - x_1)}{\sqrt{1 + (\cos^2 \pi/16)[(x_2 - x_1)/0.2]^2}} = 0$$

$$\frac{(\cos^2 \pi/16)(x_2 - x_1)}{\sqrt{1 + (\cos^2 \pi/16)[(x_2 - x_1)/0.2]^2}} - \frac{(\cos^2 \pi/8)(x_3 - x_2)}{\sqrt{1 + (\cos^2 \pi/8)[(x_3 - x_2)/0.2]^2}} = 0$$

$$\frac{(\cos^2 \pi/8)(x_3 - x_2)}{\sqrt{1 + (\cos^2 \pi/8)[(x_3 - x_2)/0.2]^2}} - \frac{(\cos^2 3\pi/16)x_3}{\sqrt{1 + (\cos^2 3\pi/16)(x_3/0.2)^2}} = 0$$

9. Repeat Exercise 8 but with $\omega = 1.8$.
10. For the following system find $x_k^{(3)}$, $k = 1, 2, 3, 4, 5, 6, 7$, by the generalized Newton's method. Do all calculations to three decimal places and use $x_k^{(1)} = 0$, $\omega = 1$.

$$\begin{aligned} 4x_1 - x_2 &= 5, \\ 4x_2 - x_3 + x_6 &= -4, \\ x_1 + 4x_3 + x_4 + x_7 &= 0, \\ x_1 + x_2 + 4x_4 + x_5 &= 6, \\ -2x_2 + 4x_5 - x_6 &= -6, \\ -x_3 + 4x_6 - x_7 &= 3, \\ x_3 + x_4 + x_5 + x_6 - 4x_7 &= 12. \end{aligned}$$

11. Without special considerations, the method of Gauss elimination applied to system (3.2) can be described as follows. Select an equation from system

(3.2) in which the coefficient of x_1, say a_{k1}, is of maximum absolute value. Then add the multiple $-a_{j1}/a_{k1}$, $j \neq k$, of this kth equation to the jth equation for each of $j = 1, 2, \ldots, k-2, k-1, k+1, k+2, \ldots, n$. Set the kth equation aside and consider the remaining $(n-1)$ equations, which contain only the $(n-1)$ unknowns $x_2, x_3, \ldots, x_n$. Select from these an equation in which the coefficient of x_2 is of maximum absolute value and add suitable multiples of this equation to the remaining $(n-2)$ equations so that in each resulting equation the coefficient of x_2 is zero. Set aside the equation whose x_2 coefficient is nonzero and consider the remaining $(n-2)$ equations in the $(n-2)$ unknowns $x_3, x_4, \ldots, x_n$. In an inductive fashion, apply the indicated elimination process until, in a finite number of steps, there results a system of equations of the form

$$
\begin{aligned}
c_{11}x_1 + c_{12}x_2 + c_{13}x_3 + \cdots + c_{1,n-1}x_{n-1} + c_{1n}x_n &= d_1,\\
c_{22}x_2 + c_{23}x_3 + \cdots + c_{2,n-1}x_{n-1} + c_{2n}x_n &= d_2,\\
c_{33}x_3 + \cdots + c_{3,n-1}x_{n-1} + c_{3n}x_n &= d_3,\\
\ddots \qquad \vdots \qquad\qquad \vdots \quad &\quad \vdots\\
c_{n-1,n-1}x_{n-1} + c_{n-1,n}x_n &= d_{n-1},\\
c_{nn}x_n &= d_n,
\end{aligned}
\tag{3.66}
$$

where $c_{ii} \neq 0$, $i = 1, 2, \ldots, n$. From the last equation of system (3.66) calculate x_n. Using this value of x_n, back-substitution into the next-to-the-last equation of system (3.66) readily yields x_{n-1}. Then using the values of x_n and x_{n-1}, back-substitution into the second-from-the-last equation of system (3.66) readily yields x_{n-2}. Proceeding then in the indicated fashion, one readily finds by back-substitution $x_{n-3}, x_{n-4}, \ldots, x_3, x_2, x_1$.

(a) Apply Gauss elimination to find the solution of the linear systems given in Exercises 1(b), 1(c), 5(d), and 10.

(b) Is there any relationship between Gauss elimination and the method of Section 3.4?

12. Without special considerations, the method of Gauss–Jordan elimination applied to system (3.2) can be described as follows. First construct system (3.66) from (3.2). Then add a suitable multiple of the second equation to eliminate the x_2 term of the first equation. Next add a suitable multiple of the third equation to eliminate the x_3 term of the new first equation and the x_3 term of the second equation. In an inductive fashion, apply the indicated elimination process until, in a finite number of steps, (3.66) is transformed into an equivalent, diagonal system, which is then solved easily.

(a) Apply Gauss–Jordan elimination to the linear systems of Exercises 1(b), 1(c), 5(d), and 10.
(b) Is there any relationship between Gauss–Jordan elimination and the method of Section 3.4?

Chapter 4

SECOND-ORDER DIFFERENTIAL AND DIFFERENCE EQUATIONS

4.1. Introduction

In applying Newton's equation

$$F = ma$$

to the simplest of dynamical problems, one is led directly to equations of the form

$$y'' = f(x,y,y'), \tag{4.1}$$

where x represents time and y represents distance. For this reason, we shall concentrate on equations of form (4.1).

If a and b are two real numbers, then by an open interval I we shall mean a set of real numbers x that satisfy either $a < x < b$, where $a < b$, or $a < x$. (Although these are not the only open intervals that can be defined, they are the only ones we shall use.) The closure $\bar{I}$ of I is the set $a \leq x \leq b$ if I is $a < x < b$, and it is the set $a \leq x$ if I is $a < x$.

Definition 4.1. Let I be an open interval of real numbers. For x in I and for all real numbers u and v, let $f(x,u,v)$ be defined and continuous. Then

$$y'' = f(x,y,y') \tag{4.2}$$

is said to be a *second-order* (*ordinary*) *differential equation* on I.

Example 1. $y'' = y\sqrt{1 - x^2} + (y')^2$ is a second-order differential equation on $-1 < x < 1$.

Example 2. $y'' = xy + (y')^3 - \sin xy$ is a second-order differential equation on $x > 0$.

Definition 4.2. If (4.2) is defined on open interval I, then $y = g(x)$ is said to be a solution of (4.2) on I if and only if

(a) $y = g(x)$ is defined and continuous on $\bar{I}$,
(b) $y = g(x)$ has continuous first- and second-order derivatives on I, and
(c) $g''(x) \equiv f(x, g(x), g'(x))$ is an *identity* in x on I.

Example 1. The function $y = x^3$, defined on $x \geq 0$, is a solution of $y'' = (3xy' - 3y)/x^2$ on $x > 0$ because $y = x^3$ is continuous on $x \geq 0$, $y' = 3x^2$ and $y'' = 6x$ are continuous on $x > 0$, and

$$6x \equiv \frac{(3x)(3x^2) - 3x^3}{x^2}, \qquad x > 0.$$

Example 2. The function $y = x^2$, defined on $x \geq 0$, is *not* a solution of $y'' + 2y' + xy = x^3$ on $x > 0$ because

$$2 + 2(2x) + x(x^2) \neq x^3, \qquad x > 0.$$

Example 3. Each of $y = \sin x$, $\cos x$, $-3 \sin x$, $4 \cos x$, $5 \sin x - 8 \cos x$, defined on $0 \leq x \leq 2\pi$, is a solution of $y'' + y = 0$ on $0 < x < 2\pi$.

With regard to (4.1), two fundamental problems are of broad physical interest. The first, called an *initial-value problem*, is one in which a particle at a fixed point P_0 is propelled in a fixed direction at a fixed initial velocity v_0. Its subsequent motion is determined by an equation of type (4.2) and one must construct the trajectory of the motion. The second, called a *boundary-value problem*, is one in which one wishes to determine the initial direction and velocity with which to propel a particle from a point P_0 so that it will arrive at a given point P_1 when its motion is determined by an equation of type (4.2). These problems are defined now formally as follows.

Definition 4.3. Let I be an open interval and let α and β be constants. The problem of constructing a solution on I of (4.2) which in addition satisfies

$$y(a) = \alpha, \qquad y'(a) = \beta \tag{4.3}$$

is called an *initial-value problem*.

Example. The problem of finding a solution of

$$y'' = (xy')^2 - e^y$$

on $x > 0$ which satisfies

$$y(0) = 0, \qquad y'(0) = 1$$

is an initial-value problem.

Definition 4.4. Let I be an open interval $a < x < b$ and let α and β be constants. The problem of finding a solution on I of (4.2) which in addition satisfies

$$y(a) = \alpha, \qquad y(b) = \beta \tag{4.4}$$

is called a *boundary-value problem.*

Example. The problem of finding a solution of

$$y'' = x^2y' - e^y$$

on $0 < x < 1$ which in addition satisfies

$$y(0) = 0, \qquad y(1) = 2$$

is a boundary-value problem.

4.2. Linear Differential Equations

Mathematicians have provided general *analytical* methods for solving second-order differential equations only for certain classes of such equations which are linear. In this and the next two sections, attention will be directed to these equations and to a survey of the methods for their solution. For linear differential equations that cannot be solved by existing analytical methods, and for nonlinear differential equations, the scientific importance of which cannot be overestimated easily, we shall develop *numerical* methods in Chapters 5, 6, and 7.

Definition 4.5. On an open interval I, let $P(x)$, $Q(x)$, $R(x)$ be continuous. Then on I the differential equation

$$y'' = -P(x)y' - Q(x)y + R(x)$$

or, equivalently,

$$y'' + P(x)y' + Q(x)y = R(x) \tag{4.5}$$

is said to be *linear.*

Examples. On $x > 0$, each of the following is a linear differential equation:

$$y'' + 5y' + 4y = 0,$$

$$y'' + 5y' + 4y = \sin x,$$

$$y'' + 5xy' + 4x^2y = \sin x,$$

$$y'' + \frac{5}{x}y' + \frac{4}{x^2}y = 0.$$

Definition 4.6. Equation (4.5) is said to be *homogeneous* if and only if $R(x) \equiv 0$ on I.

Basic to the study of (4.5) is the concept of linearly independent functions, which can be formulated quite simply as follows.

Definition 4.7. On the closure $\bar{I}$ of open interval I, two functions $y_1(x)$ and $y_2(x)$ are *linearly independent* if and only if neither one is a constant multiple of the other.

Examples. On $0 \leq x \leq 1$,

x, x^2	are linearly independent,
x^2, x^3	are linearly independent,
$\sin x, \cos x$	are linearly independent,
$x, -5x$	are not linearly independent,
$0, x^3$	are not linearly independent,
x, e^x	are linearly independent,
$1, e^x$	are linearly independent.

Then the following fundamental theorems are well known [see, e.g., Coddington and Levinson or Greenspan (1)].

Theorem 4.1. On a given open interval I, let

$$y'' + P(x)y' + Q(x)y = 0 \tag{4.6}$$

be a linear differential equation. Then on $\bar{I}$ there exists a pair of linearly independent functions $y_1(x)$, $y_2(x)$, each of which is a solution of (4.6) on I. Moreover, for any constant values c_1 and c_2, the function

$$y(x) = c_1 y_1(x) + c_2 y_2(x) \tag{4.7}$$

is also a solution of (4.6), and, indeed, *every* solution of (4.6) can be derived from (4.7) by an appropriate choice of constants c_1 and c_2.

Example. For $x > 0$, consider

$$y'' + y = 0.$$

Then $y_1 = \sin x$, $y_2 = \cos x$ are solutions that are linearly independent on $x \geq 0$. Therefore,

$$y(x) = c_1 \sin x + c_2 \cos x$$

is also a solution for any constant values c_1 and c_2, and every solution is of this form.

By virtue of the results stated in Theorem 4.1, (4.7) is called the *general solution* of (4.6).

Theorem 4.2. On a given open interval I, let

$$y'' + P(x)y' + Q(x)y = R(x) \tag{4.8}$$

be a linear differential equation. Let $\bar{y}(x)$ be any solution of (4.8) and let $y_1(x)$, $y_2(x)$ be as defined in Theorem 4.1. Then, for any constant values c_1 and c_2,

$$y(x) = c_1y_1(x) + c_2y_2(x) + \bar{y}(x) \tag{4.9}$$

is a solution of (4.8). Moreover, every solution of (4.8) can be derived from (4.9) by an appropriate choice of constants c_1 and c_2.

Example. For $x > 0$, consider

$$y'' + y = 2e^x. \tag{4.10}$$

Then $\bar{y} = e^x$ is a solution of (4.10), as is

$$y = c_1 \sin x + c_2 \cos x + e^x \tag{4.11}$$

for any constants c_1 and c_2. Moreover, every solution of (4.10) is of the form (4.11).

By virtue of the results stated in Theorem 4.2, (4.9) is called the *general solution* of (4.8).

4.3. Linear Differential Equations with Constant Coefficients

In attempting to implement the results of Theorems 4.1 and 4.2 by actually seeking to construct solutions, it is natural to examine first the very simple linear differential equation

$$y'' + py' + qy = 0, \tag{4.12}$$

where p and q are real constants. For such an equation, one can even unashamedly guess solutions that are elementary functions. The guess $y = e^{mx}$ leads, when substituted into (4.12), to the identity

$$e^{mx}(m^2 + pm + q) \equiv 0. \tag{4.13}$$

But (4.13) is identically zero if and only if

$$m^2 + pm + q = 0. \tag{4.14}$$

If the two roots of (4.14) are m_1 and m_2, then (4.12) can always be solved completely by the following well-known theorem [see, e.g., Greenspan (1)].

Theorem 4.3. If m_1 and m_2 are the roots of (4.14), then on any open interval I:

(a) m_1, m_2 real and distinct yield the general solution

$$y(x) = c_1 e^{m_1 x} + c_2 e^{m_2 x} \tag{4.15}$$

of (4.12);

(b) m_1, m_2 real and equal yield the general solution

$$y(x) = c_1 e^{m_1 x} + c_2 x e^{m_1 x} \tag{4.16}$$

of (4.12); and

(c) m_1, m_2 complex conjugate, say $m_1 = \alpha + \beta i$, $m_2 = \alpha - \beta i$, $\beta \neq 0$, yield the general solution

$$y(x) = c_1 e^{\alpha x} \sin \beta x + c_2 e^{\alpha x} \cos \beta x \tag{4.17}$$

of (4.12). Moreover, as cases (a), (b), and (c) treat all possible combinations of m_1 and m_2, the general solution of (4.12) can always be found and is, indeed, one of the three forms (4.15) through (4.17).

Example 1. On any interval I, the general solution of

$$y'' + 4y' + 3y = 0$$

is

$$y(x) = c_1 e^{-x} + c_2 e^{-3x}$$

because the roots $m_1 = -1$, $m_2 = -3$ of

$$m^2 + 4m + 3 = 0$$

are real and unequal.

Example 2. On any interval I, the general solution of

$$y'' + 4y' + 4y = 0$$

is

$$y(x) = c_1 e^{-2x} + c_2 x e^{-2x}$$

because the roots $m_1 = -2$, $m_2 = -2$ of

$$m^2 + 4m + 4 = 0$$

are real and equal.

Example 3. On any interval I, the general solution of

$$y'' + y = 0$$

is

$$y(x) = c_1 \sin x + c_2 \cos x$$

because the roots $m_1 = i, m_2 = -i$ of

$$m^2 + 1 = 0$$

are conjugate complex.

Next, consider an equation that is slightly more complicated in structure than (4.12), that is, an equation of the form

$$y'' + py' + qy = R(x), \tag{4.18}$$

where p and q are constants and $R(x)$ is continuous on a given interval I. Although a particular solution of (4.18) is always known to exist [see, e.g., Hildebrand (1)], the method of undetermined coefficients, as illustrated in the next examples, is exceptionally useful in practice.

Example 4. Find the general solution on $0 < x < 1$ of

$$y'' - 5y' + 4y = x^2 - 2x + 1. \tag{4.19}$$

Solution. Let

$$\bar{y}(x) = ax^2 + bx + c \tag{4.20}$$

be a particular solution of (4.19). Then

$$\bar{y}'(x) = 2ax + b,$$

$$\bar{y}''(x) = 2a,$$

and substitution into (4.19) implies that

$$2a - 5(2ax + b) + 4(ax^2 + bx + c) \equiv x^2 - 2x + 1$$

or, equivalently,

$$4ax^2 + (4b - 10a)x + (4c - 5b + 2a) \equiv x^2 - 2x + 1. \tag{4.21}$$

Setting corresponding coefficients equal in (4.21) implies that

$$a = \tfrac{1}{4}, \qquad b = \tfrac{1}{8}, \qquad c = \tfrac{9}{32},$$

so that

$$\bar{y}(x) = \tfrac{1}{4}x^2 + \tfrac{1}{8}x + \tfrac{9}{32}.$$

Since the general solution of

$$y'' - 5y' + 4y = 0$$

is

$$y(x) = c_1 e^x + c_2 e^{4x},$$

it follows from Theorem 3.2 that the general solution of (4.19) on $0 < x < 1$ is

$$y(x) = c_1 e^x + c_2 e^{4x} + \tfrac{1}{4}x^2 + \tfrac{1}{8}x + \tfrac{9}{32}.$$

Example 5. On $0 < x < 1$, find the general solution of

$$y'' - 5y' + 4y = \sin x. \tag{4.22}$$

Solution. Let

$$\bar{y}(x) = a \sin x + b \cos x.$$

Then

$$\bar{y}'(x) = a \cos x - b \sin x,$$

$$\bar{y}''(x) = -a \sin x - b \cos x.$$

Substitution into (4.22) implies

$$(-a \sin x - b \cos x) - 5(a \cos x - b \sin x) + 4(a \sin x + b \cos x) \equiv \sin x$$

or, equivalently,

$$(3a + 5b) \sin x + (3b - 5a) \cos x \equiv \sin x.$$

Setting

$$3a + 5b = 1,$$

$$3b - 5a = 0$$

implies that $a = \frac{3}{34}$, $b = \frac{5}{34}$, so

$$\bar{y}(x) = \tfrac{3}{34} \sin x + \tfrac{5}{34} \cos x,$$

and it follows readily that the general solution of (4.22) on $0 < x < 1$ is

$$y(x) = c_1 e^x + c_2 e^{4x} + \tfrac{3}{34} \sin x + \tfrac{5}{34} \cos x.$$

4.4. Linear Differential Equations with Nonconstant Coefficients

The final step, that of considering p and q in (4.18) to be functions of x, leads back to (4.5). Extensive study of such equations has been made only under the assumptions that $P(x)$ and $Q(x)$ are *rational* functions. For this class of equations the method of undetermined coefficients, discussed in Section 4.3, combined with infinite series has been of exceptional value. One must always be aware, however, that reorganization of, or differentiation of, terms of a series are sensitive procedures that are not always allowable, although theoretical support for each step in the examples of this section can be found in Bieberbach. The general series technique that is recommended is called the method of Frobenius and is illustrated in the following example.

Example 1. On $0 < x < 1$, find the general solution of

$$y'' - 4xy' + (4x^2 - 2)y = 0. \tag{4.23}$$

Solution. Let

$$y = \sum_{n=0}^{\infty} a_n x^{n+s}, \tag{4.24}$$

where $a_0 = 1$ and s is a parameter to be determined. Then

$$y' = \sum_{n=0}^{\infty} a_n(n + s)x^{n+s-1}, \tag{4.25}$$

$$y'' = \sum_{n=0}^{\infty} a_n(n + s)(n + s - 1)x^{n+s-2}. \tag{4.26}$$

Substitution of (4.24) through (4.26) into (4.23) implies that

$$\sum_{n=0}^{\infty} a_n(n + s)(n + s - 1)x^{n+s-2} - \sum_{n=0}^{\infty} 4a_n(n + s)x^{n+s} + \sum_{n=0}^{\infty} 4a_n x^{n+s+2} - \sum_{n=0}^{\infty} 2a_n x^{n+s} \equiv 0,$$

from which it follows that

$$\sum_{n=-2}^{\infty} a_{n+2}(n + 2 + s)(n + 1 + s)x^{n+s} - \sum_{n=0}^{\infty} 4a_n(n + s)x^{n+s} + \sum_{n=2}^{\infty} 4a_{n-2}x^{n+s} - \sum_{n=0}^{\infty} 2a_n x^{n+s} \equiv 0.$$

Reorganization of the terms of the above series and recalling that $a_0 = 1$ implies that

$$\begin{aligned} s(s - 1)x^{s-2} &+ a_1(s + 1)sx^{s-1} + [a_2(s + 2)(s + 1) - 4s - 2]x^s \\ &+ [a_3(s + 3)(s + 2) - 4a_1(s + 1) - 2a_1]x^{s+1} \\ &+ \sum_{n=2}^{\infty} x^{n+s}[a_{n+2}(n + s + 2)(n + s + 1) \\ &- 4a_n(n + s) + 4a_{n-2} - 2a_n] \equiv 0. \end{aligned} \tag{4.27}$$

From (4.27), then, setting corresponding coefficients equal yields

(4.28) $$s(s - 1) = 0,$$

(4.29) $$a_1(s + 1)s = 0,$$

(4.30) $$a_2(s + 2)(s + 1) - 4s - 2 = 0,$$

(4.31) $$a_3(s + 3)(s + 2) - 4a_1(s + 1) - 2a_1 = 0,$$

(4.32) $$a_{n+2}(n + s + 2)(n + s + 1) - 4a_n(n + s) + 4a_{n-2} - 2a_n = 0, \quad n = 2, 3, \ldots.$$

From (4.28), we must consider the two cases

(4.33) $$s = 0, \qquad s = 1.$$

Consider first $s = 0$. Then (4.29) through (4.32) imply that

(4.34) $$a_1 \cdot 0 = 0,$$

(4.35) $$a_2 = 1,$$

(4.36) $$a_3 = a_1,$$

(4.37) $$a_{n+2} = \frac{2a_n(1 + 2n) - 4a_{n-2}}{(n + 2)(n + 1)}, \qquad n = 2, 3, \ldots.$$

From (4.34), a_1 is arbitrary. When this is the case, it is very often convenient to let the arbitrary constant be zero. Hence set $a_1 = 0$. Then (4.36) implies that $a_3 = 0$ and (4.37) implies that $a_5 = a_7 = a_9 = \cdots = 0$. For $n = 2, 4, 6, \ldots$, (4.37) implies that

$$a_4 = \frac{2a_2(1 + 4) - 4}{4 \cdot 3} = \frac{1}{2}, a_6 = \frac{1}{3!}, a_8 = \frac{1}{4!}, \ldots, a_{2n} = \frac{1}{n!}.$$

Thus, for $s = 0$, (4.24) yields

(4.38) $$y_1(x) = 1 + x^2 + \frac{x^4}{2!} + \frac{x^6}{3!} + \frac{x^8}{4!} + \cdots + \frac{x^{2n}}{n!} + \cdots \equiv e^{x^2}.$$

For $s = 1$, (4.29) through (4.32) imply that

(4.39) $$2a_1 = 0,$$

(4.40) $$a_2 = 1,$$

(4.41) $$6a_3 = 5a_1,$$

(4.42) $$a_{n+2} = \frac{2a_n(3 + 2n) - 4a_{n-2}}{(n + 3)(n + 2)}, \qquad n = 2, 3, \ldots.$$

From (4.39), (4.41), and (4.42), it follows that $a_n = 0$, for n odd. From (4.42), for n even one has

$$a_4 = \frac{2 \cdot a_2(3 + 4) - 4}{5 \cdot 4} = \frac{1}{2}, a_6 = \frac{1}{3!}, a_8 = \frac{1}{4!}, \ldots, a_{2n} = \frac{1}{n!}.$$

For $s = 1$, then, (4.24) yields

$$y_2(x) = x + x^3 + \frac{x^5}{2!} + \frac{x^7}{3!} + \cdots + \frac{x^{2n+1}}{n!} + \cdots \equiv xe^{x^2}. \tag{4.43}$$

The reader unfamiliar with the theory of series manipulations can, and indeed should, check now that $y_1(x)$ and $y_2(x)$ are linearly independent solutions of (4.23), the general solution of which is

$$y(x) = c_1 e^{x^2} + c_2 x e^{x^2}.$$

The method of Frobenius, as illustrated in Example 1, applies on certain restricted intervals (Bieberbach) to the class of interesting equations of the type

$$y'' + \frac{(\alpha + \beta x^m)}{x(1 + \gamma x^m)} y' + \frac{(\delta + \rho x^m)}{x^2(1 + \gamma x^m)} y = 0, \tag{4.44}$$

where α, β, γ, δ, and ρ are constants. The following two examples illustrate how this can be done.

Example 2. On $\varepsilon < x < 1$, where $0 < \varepsilon < 1$, find the general solution of

$$y'' + \frac{1}{x} y' + \left(1 - \frac{1}{4x^2}\right) y = 0. \tag{4.45}$$

Solution. Substitution of (4.24) through (4.26) into (4.45) and recombination under the assumption that $a_0 = 1$ yields

$$(s + \tfrac{1}{2})(s - \tfrac{1}{2})x^s + [(s + 1)^2 - \tfrac{1}{4}]a_1 x^{s+1}$$
$$+ \sum_{n=2}^{\infty} [\{[(n + s)^2 - \tfrac{1}{4}]a_n + a_{n-2}\} x^{n+s}] \equiv 0.$$

Setting corresponding coefficients equal to zero implies that

$$(s + \tfrac{1}{2})(s - \tfrac{1}{2}) = 0, \tag{4.46}$$

$$[(s + 1)^2 - \tfrac{1}{4}]a_1 = 0, \tag{4.47}$$

$$a_n[(n + s)^2 - \tfrac{1}{4}] = -a_{n-2}, \qquad n = 2, 3, \ldots, \tag{4.48}$$

and, from (4.46), we consider the two parameter values $s = \frac{1}{2}$, $s = -\frac{1}{2}$.

For $s = \frac{1}{2}$, (4.47) and (4.48) become

$$2a_1 = 0, \tag{4.49}$$

$$a_n = -\frac{a_{n-2}}{n(n+1)}, \qquad n = 2, 3, \ldots. \tag{4.50}$$

Thus, for n odd, (4.49) and (4.50) imply that $a_n = 0$, while for n even one has

$$a_2 = -\frac{1}{3!}, a_4 = \frac{1}{5!}, \ldots, a_n = \frac{(-1)^{n/2}}{(n+1)!}.$$

Hence (4.24) yields

$$y_1(x) = x^{1/2} - \frac{x^{5/2}}{3!} + \frac{x^{9/2}}{5!} - \frac{x^{13/2}}{7!} \cdots \equiv x^{-1/2} \sin x.$$

For $s = -\frac{1}{2}$, (4.47) and (4.48) become

$$0 \cdot a_1 = 0,$$

$$a_n = -\frac{a_{n-2}}{n(n-1)}, \qquad n = 2, 3, \ldots.$$

Since a_1 is arbitrary, let it be zero, so that $a_n = 0$ for n odd. For n even one has

$$a_2 = -\frac{1}{2!}, a_4 = \frac{1}{4!}, a_6 = -\frac{1}{6!}, \ldots, a_n = \frac{(-1)^{n/2}}{n!},$$

and (4.24) yields

$$y_2(x) = x^{-1/2} - \frac{x^{3/2}}{2!} + \frac{x^{7/2}}{4!} - \frac{x^{11/2}}{6!} + \cdots \equiv x^{-1/2} \cos x.$$

The general solution then is

$$y(x) = c_1 x^{-1/2} \sin x + c_2 x^{-1/2} \cos x.$$

Example 3. Find the general solution on $\varepsilon < x < 1$, where $0 < \varepsilon < 1$, of

$$y'' - y' - \frac{1}{x} y = 0. \tag{4.51}$$

Solution. Substitution of (4.24) through (4.26) into (4.51) and recombination under the assumption that $a_0 = 1$ yields

$$s(s-1)x^{s-2} + \sum_{n=0}^{\infty} [(n+s+1)(n+s)a_{n+1} - (n+s+1)a_n]x^{n+s+1} \equiv 0,$$

so

$$s(s-1)=0, \tag{4.52}$$

$$(n+s+1)(n+s)a_{n+1}=(n+s+1)a_n, \qquad n=0,1,\ldots. \tag{4.53}$$

Note that the factor $(n+s+1)$ in (4.53) must not be divided out until one knows that it is *not* zero.

From (4.52), one must consider the two cases

$$s=0, \qquad s=1.$$

The case $s=0$ implies from (4.53) that

$$(n+1)(n)a_{n+1}=(n+1)a_n, \qquad n=0,1,2,\ldots,$$

or that

$$a_{n+1}=\frac{a_n}{n}.$$

But, for $n=0$, $a_1=1/0$, which is not defined, and the method fails.

For $s=1$, the method yields, in the straightforward way, the solution

$$y_1(x)=xe^x.$$

When the two parameter values for s do not yield two solutions $y_1(x)$ and $y_2(x)$, as in the present case, it is known [Greenspan (1)] that $y_2(x)$ can be obtained by undetermined coefficients from

$$y_2=y_1(x)\log x+\sum_{n=0}^{\infty}b_nx^{n+\bar{s}},$$

where $\bar{s}$ is the smaller of the two parameter values s. In the present example, then,

$$y_2=xe^x\log x+\sum_{n=0}^{\infty}b_nx^n. \tag{4.54}$$

From (4.54)

$$y_2'=e^x\log x+xe^x\log x+e^x+\sum_{n=0}^{\infty}nb_nx^{n-1},$$

$$y_2''=2e^x\log x+\frac{e^x}{x}+xe^x\log x+2e^x+\sum_{n=2}^{\infty}(n)(n-1)b_nx^{n-2},$$

and substitution into the differential equation yields

$$e^x+\frac{e^x}{x}+\sum_{n=2}^{\infty}n(n-1)b_nx^{n-2}-\sum_{n=1}^{\infty}nb_nx^{n-1}-\sum_{n=0}^{\infty}b_nx^{n-1}\equiv 0. \tag{4.55}$$

Multiplication by x, substitution of the series expansion for e^x, and recombination reduces (4.55) further to

$$1 - b_0 + \sum_{n=1}^{\infty} x^n \left[\frac{n+1}{n!} + (n)(n+1)b_{n+1} - (n+1)b_n \right] \equiv 0,$$

so

(4.56) $$1 - b_0 = 0,$$

(4.57) $$\frac{n+1}{n!} + n(n+1)b_{n+1} - (n+1)b_n \equiv 0, \qquad n = 1, 2, \ldots.$$

Thus

(4.58) $$b_0 = 1$$

and, since $n + 1 \neq 0$,

(4.59) $$b_{n+1} = \frac{b_n}{n} - \frac{1}{n(n!)}, \qquad n = 1, 2, \ldots.$$

From (4.59) it follows that b_1 is arbitrary, so, let

$$b_1 = 0.$$

Then

$$b_2 = -\frac{1}{1(1)!} = -1,$$

$$b_3 = -\frac{1}{2} - \frac{1}{2(2!)} = -\frac{1}{2!}\left(1 + \frac{1}{2}\right),$$

$$b_4 = -\frac{1}{3!}\left(1 + \frac{1}{2}\right) - \frac{1}{3(3!)} = -\frac{1}{3!}\left(1 + \frac{1}{2} + \frac{1}{3}\right),$$

$$\vdots$$

$$b_k = -\frac{1}{(k-1)!}\left(1 + \frac{1}{2} + \cdots + \frac{1}{k-1}\right).$$

Thus

$$y_2(x) = xe^x \log x + 1 - \sum_{n=2}^{\infty} \left[\left(1 + \frac{1}{2} + \frac{1}{3} + \cdots + \frac{1}{n-1}\right) \frac{x^n}{(n-1)!} \right],$$

and the general solution of (4.51) is

$$y = c_1 xe^x + c_2 \left\{ xe^x \log x + 1 - \sum_{n=2}^{\infty} \left[\left(1 + \frac{1}{2} + \frac{1}{3} + \cdots + \frac{1}{n-1}\right) \frac{x^n}{(n-1)!} \right] \right\}.$$

It is to be observed that, *more often than not*, the method of Frobenius results in series solutions that cannot be expressed in simpler form. The evaluation of these series for fixed values of x was at one time considered to be so important that major organizations such as the National Bureau of Standards devoted large numbers of man-hours during the 1930s to constructing tables of such values. This work, done originally on desk calculators, was redone recently with exceptional speed on modern digital computers. The resulting tables are called tables of special functions, because any solution of (4.44) is called a *special function.* Examples of such functions are Bessell functions, Legendre functions, and Hermite polynomials, which are the respective solutions generated by the method of Frobenius of

$$y'' + \frac{1}{x}y' + \left(1 - \frac{v^2}{x^2}\right)y = 0, \qquad v \text{ a constant,}$$

$$y'' - \frac{2x}{1 - x^2}y' + \frac{v(v + 1)}{1 - x^2}y = 0, \qquad v \text{ a constant,}$$

$$y'' - 2xy' + 2ny = 0, \qquad n \text{ an integer.}$$

4.5. Initial- and Boundary-Value Problems

After one has obtained the general solution of (4.5), one has a function with two arbitrary constants. It is important to ask whether the two additional conditions (4.3), or (4.4), determine these constants uniquely. Although this is known to be correct for (4.3), it is not correct for (4.4); that is, *initial-value problems for* (4.5) *have unique solutions* [Greenspan (1)], whereas boundary-value problems need not. Indeed, on $0 < x < 2\pi$ consider the differential equation

$$y'' + y = 0, \tag{4.60}$$

the general solution of which is

$$y(x) = c_1 \sin x + c_2 \cos x. \tag{4.61}$$

Given the boundary conditions

$$y(0) = 0, \qquad y(2\pi) = 0, \tag{4.62}$$

then (4.61) and (4.62) imply that

$$\begin{aligned} 0 &= c_1 \sin 0 + c_2 \cos 0, \\ 0 &= c_1 \sin 2\pi + c_2 \cos 2\pi, \end{aligned} \tag{4.63}$$

from which one has that c_1 is arbitrary while $c_2 = 0$. Thus

$$y = c_1 \sin x \tag{4.64}$$

constitutes an infinite number of solutions that satisfy the boundary-value problem defined by (4.60) and (4.62).

The following sufficiency condition, which is valid in many applied problems, is useful for establishing the existence and uniqueness of solutions of boundary-value problems.

Theorem 4.4. The boundary-value problem defined by (4.4) and (4.5) has a unique solution if $Q \leq 0$ on I.

For the proof, see Keller.

For completeness, consider finally two examples that illustrate the solution of initial- and boundary-value problems for linear differential equations.

Example 1. On $0 < x < 1$, consider (4.19). Then find the unique solution of (4.19) which satisfies the initial conditions $y(0) = y'(0) = 0$.

Solution. Since the general solution is

$$y(x) = c_1 e^x + c_2 e^{4x} + \tfrac{1}{4}x^2 + \tfrac{1}{8}x + \tfrac{9}{32},$$

then

$$y'(x) = c_1 e^x + 4c_2 e^{4x} + \frac{x}{2} + \frac{1}{8}.$$

Hence the initial conditions imply that

$$0 = c_1 + c_2 + \tfrac{9}{32},$$

$$0 = c_1 + 4c_2 + \tfrac{1}{8},$$

the solution of which is $c_1 = -\tfrac{1}{3}$, $c_2 = \tfrac{5}{96}$. Thus the unique solution is

$$y = -\tfrac{1}{3}e^x + \tfrac{5}{96}e^{4x} + \tfrac{1}{4}x^2 + \tfrac{1}{8}x + \tfrac{9}{32}.$$

Example 2. For $0.0001 < x < 0.09$, consider (4.45). Then find the unique solution of (4.45) which satisfies the boundary conditions $y(0.0001) = 0$, $y(0.09) = 1$.

Solution. For (4.45), $Q(x) \equiv (1 - 1/(4x^2)) \leq 0$ when $0.0001 < x < 0.09$. Thus Theorem 4.4 is applicable and we are assured a priori that the solution exists and is unique. Since the general solution of (4.45) is

$$y(x) = c_1 x^{-1/2} \sin x + c_2 x^{-1/2} \cos x, \tag{4.65}$$

the given boundary conditions imply that

$$0 = c_1\frac{\sin(0.0001)}{0.01} + c_2\frac{\cos(0.0001)}{0.01},$$

$$1 = c_1\frac{\sin(0.09)}{0.3} + c_2\frac{\cos(0.09)}{0.3}.$$

The unique solution of the above equations is

$$c_1 = \frac{-(0.3)\cos(0.0001)}{\sin(0.0001)\cos(0.09) - \sin(0.09)\cos(0.0001)},$$

$$c_2 = \frac{(0.3)\sin(0.0001)}{\sin(0.0001)\cos(0.09) - \sin(0.09)\cos(0.0001)},$$

which then yield the desired solution from (4.65).

4.6. Difference Equations

In the same fashion that differentiable functions are related to differential equations, so discrete functions are related to difference equations, and in this section we shall develop some of the corresponding analogies.

Definition 4.8. Let $x_0, x_1, \ldots, x_n$ be an R_{n+1} set. Let the interior of R_{n+1}, denoted by $I(R_{n+1})$, be the set $x_1, x_2, \ldots, x_{n-1}$. Then an equation of the form

$$y_{i+1} = F(x_i, y_{i-1}, y_i, \Delta x), \qquad i = 1, 2, \ldots, n-1, \tag{4.66}$$

where, in fact, F need not depend on x_i, y_i, or Δx but must depend on y_{i-1}, is called a *second-order difference equation* on $I(R_{n+1})$.

Example. Let R_5 be the set $x_0 = 0$, $x_1 = 1$, $x_2 = 2$, $x_3 = 3$, $x_4 = 4$. On $I(R_5)$ each of the following is a second-order difference equation:

$$y_{i+1} = y_i + 2y_{i-1} + 2(\Delta x)x_i, \qquad i = 1, 2, 3,$$

$$y_{i+1} = y_{i-1}(y_i)^2, \qquad i = 1, 2, 3,$$

$$y_{i+1} = (\Delta x)e^{y_{i-1}}, \qquad i = 1, 2, 3.$$

Second-order difference equations arise in a variety of natural ways. Suppose, for example, that in the *nonlinear* differential equation

$$y'' - (y')^2 = x \tag{4.67}$$

one substitutes the three-point central-difference approximation (2.54) for

y'' and two-point backward-difference approximation (2.38) for y', so that

$$\frac{y_{i+1} - 2y_i + y_{i-1}}{(\Delta x)^2} - \left(\frac{y_i - y_{i-1}}{\Delta x}\right)^2 = x_i$$

or, equivalently,

$$y_{i+1} = 2y_i - y_{i-1} + (y_i - y_{i-1})^2 + x_i(\Delta x)^2, \tag{4.68}$$

which is exactly of the form (4.66). Thus, if one wishes to approximate a differential equation by replacing derivatives by differences, one arrives at equations such as (4.66). Also, it is worth noting that such substitutions replace the analytic nonlinearities in a differential equation by algebraic nonlinearities, which is of great advantage, for algebraic nonlinearities, such as y_i^2, $(y_{i-1})^2$, and y_i^7, are precisely the kinds of expressions with which one can deal effectively on a computer.

If Δx, $x_0, x_1, \ldots, x_n$ are fixed, then one can use (4.66) to generate from given y_0 and y_1, in a recursive fashion, $y_2, y_3, \ldots, y_n$, so that in fact (4.66) is a recursion formula. The problem of constructing $y_2, y_3, \ldots, y_n$ in such a fashion from given y_0 and y_1 is called an initial-value problem for (4.66). But note that given y_0 and $(y_1 - y_0)/\Delta x$, which corresponds to being given y_0 and y'_0 for $y'' = f(x,y,y')$, then one can use these, as shown in the example below, to determine y_0 and y_1. Initial-value problems for (4.66) are then usually formulated as follows.

Definition 4.9. On the R_{n+1} set $x_0, x_1, \ldots, x_n$, the problem of finding, from prescribed values for y_0 and $(y_1 - y_0)/\Delta x$, values $y_0, y_1, y_2, \ldots, y_n$ which satisfy (4.66) for $i = 1, 2, \ldots, n-1$ is called an *initial-value problem*. Any set $y_0, y_1, \ldots, y_n$ that satisfies the given initial conditions and the given difference equation on $I(R_{n+1})$ is called a *solution* of the problem.

Example. The problem of finding values y_0, y_1, y_2, y_3, y_4 on the R_5 set $x_0 = 0, x_1 = 1, x_2 = 2, x_3 = 3, x_4 = 4$ such that

$$y_0 = 1, \tag{4.69}$$

$$\frac{y_1 - y_0}{\Delta x} = 2, \tag{4.70}$$

$$y_{i+1} = 5(y_{i-1})^2 - (y_i)^2, \qquad i = 1, 2, 3, \tag{4.71}$$

is an initial-value problem. From (4.69) and (4.70), $y_1 = 3$, so, from (4.71),

$$y_2 = 5(1)^2 - (3)^2 = -4,$$

$$y_3 = 5(3)^2 - (-4)^2 = 29,$$

$$y_4 = 5(-4)^2 - (29)^2 = -761.$$

Thus a solution of initial-value problem (4.69) through (4.71) is $y_0 = 1$, $y_1 = 3$, $y_2 = -4$, $y_3 = 29$, $y_4 = -761$.

The following theorem is now proved easily.

Theorem 4.5. If F in (4.66) is always defined, then the solution of an initial-value problem for (4.66) always exists and is unique.

Proof. If one is given

$$y_0 = \alpha, \qquad \frac{y_1 - y_0}{\Delta x} = \beta,$$

then

$$y_0 = \alpha, \qquad y_1 = y_0 + \beta\,\Delta x.$$

Using these values of y_0 and y_1, the existence and uniqueness of the solution follow immediately from the recursive structure of (4.66).

Note that the existence and uniqueness theorem, above, is obtained with the greatest of ease and indeed is completely arithmetic in nature. Such theorems for differential equations are exceptionally difficult to obtain and, when obtainable, often require the most advanced topological concepts and techniques in their proofs. Also, it is worth noting that many mathematicians claim that nonlinear difference equations are more difficult to solve than nonlinear differential equations, so that Theorem 4.5 seems to be somewhat paradoxical. However, the enigma is resolved quite easily, for concern has been directed primarily toward the construction of *continuous* solutions of difference equations (see, e.g., Levy and Lessman), whereas the approach here is concerned, more appropriately, with the construction of *discrete* functions which are solutions.

Definition 4.10. On the R_{n+1} set $x_0, x_1, \ldots, x_n$, the problem of finding $y_0, y_1, \ldots, y_n$ which have prescribed values for y_0 and y_n and which satisfy (4.66) for $i = 1, 2, \ldots, n-1$ is called a *boundary-value problem.* Any set $y_0, y_1, \ldots, y_n$ which has the given boundary values for y_0 and y_n and satisfies the given difference equation on $I(R_{n+1})$ is called a *solution* of the problem.

Example 1. The problem of finding y_0, y_1, y_2, y_3, y_4 on the R_5 set $x_0 = 0$, $x_1 = 1$, $x_2 = 2$, $x_3 = 3$, $x_4 = 4$ such that

(4.72) $$y_0 = 0,$$

(4.73) $$y_4 = 8,$$

(4.74) $$y_{i+1} = 2y_i - y_{i-1}, \qquad i = 1, 2, 3,$$

is a boundary-value problem. For (4.74), one has for each of $i = 1, 2, 3$,

$$y_2 = 2y_1 - y_0,$$

$$y_3 = 2y_2 - y_1,$$

$$y_4 = 2y_3 - y_2,$$

which, from (4.72) and (4.73), is equivalent to the tridiagonal system

$$\begin{aligned} -2y_1 + y_2 \phantom{{}+y_3} &= 0, \\ y_1 - 2y_2 + y_3 &= 0, \\ y_2 - 2y_3 &= -8. \end{aligned}$$

The unique solution of this system is $y_1 = 2, y_2 = 4, y_3 = 6$, so the problem has the unique solution $y_0 = 0$, $y_1 = 2$, $y_2 = 4$, $y_3 = 6$, $y_4 = 8$.

As in the cases of differential equations, boundary-value problems for difference equations need not have unique solutions.

Example 2. On the R_5 set $x_0 = 0, x_1 = 1, x_2 = 2, x_3 = 3, x_4 = 4$, consider the boundary-value problem defined by

$$y_0 = 0, \tag{4.75}$$

$$y_4 = 0, \tag{4.76}$$

$$y_{i+1} = -2y_i - 2y_{i-1}, \qquad i = 1, 2, 3. \tag{4.77}$$

Then

$$y_2 = -2y_1 - 2y_0,$$

$$y_3 = -2y_2 - 2y_1,$$

$$y_4 = -2y_3 - 2y_2,$$

which, from (4.75) and (4.76), yield the tridiagonal system

$$\begin{aligned} 2y_1 + y_2 \phantom{{}+y_3} &= 0, \\ 2y_1 + 2y_2 + y_3 &= 0, \\ 2y_2 + 2y_3 &= 0. \end{aligned}$$

But the determinant of this system is zero, so there exist an infinite number of solutions.

When considering boundary-value problems in the future, then, we shall have to consider carefully the question of existence and uniqueness of solutions, and, in this connection, Theorem 3.1 can be of exceptional value.

EXERCISES

1. Determine, in each of the following, whether the given function, which is defined on the closure $\bar{I}$ of the given interval, is or is not a solution of the accompanying differential equation:
   (a) $y = -e^x((x^2/2) + x + 1)$, $y'' - 3y' + 2y = xe^x$, $-1 < x < 1$
   (b) $y = 1/x$, $x^3y'' - x^2y' - 3xy = 0, 0.1 < x$
   (c) $y = x^2 - 2x$, $y'' + y' + y = 1$, $-2 < x < 2$
2. Show that the following pairs of functions, which are defined for all x, are linearly independent on the closure $\bar{I}$ of any open interval I:
   (a) $1, x^3$
   (b) $x - 2, x - 3$
   (c) $x, \cos x$
   (d) e^x, e^{2x}
   (e) $x^n, x^{n-1}, n > 1$
   (f) $x^n e^{mx}, x^{n-1}e^{mx}, n > 1$
3. Find the general solution of each of the following and check your answers:
   (a) $y'' + 8y' + 15y = 0$
   (b) $y'' + 2y' - 15y = 0$
   (c) $y'' - 6y' + 9y = 0$
   (d) $y'' + 6y' + 9y = 0$
   (e) $y'' - 3y' + 2y = 0$
   (f) $y'' - 4y' + 13y = 0$
   (g) $2y'' + 3y' = 0$
   (h) $y'' + 25y = 0$
   (i) $4y'' + y' + y = 0$
   (j) $y'' = 0$
4. Find the general solution of each of the following and check your answers:
   (a) $y'' - 4y' + 3y = 1$
   (b) $y'' + y' - 2y = 2(1 + x - x^2)$
   (c) $y'' + y = x^3 + x$
   (d) $y'' - 6y' + 9y = e^{2x}$
   (e) $y'' - y = \sin x$
   (f) $y'' + y = x^3 - e^x - xe^x - \cos x$
5. If $Q(x) \neq 0$, show that the transformation $y = [1/Q(x)](dz/dx)(1/z)$ transforms the Riccati equation $y' + P(x)y + Q(x)y^2 = R(x)$ into $z'' + (P - (1/Q)Q')z' - RQz = 0$.
6. Find the general solution of each of the following by the method of Frobenius.
   (a) $y'' + y = 0$
   (b) $y'' + 2y' + y = 0$
   (c) $y'' + \dfrac{2}{x}y' + y = 0$
   (d) $y'' + \dfrac{1}{2x}y' + \dfrac{1}{4x}y = 0$
   (e) $y'' + \left(\dfrac{1}{2x} - 1\right)y' - \left(\dfrac{1}{2x}\right)y = 0$
   (f) $y'' - \dfrac{3}{x(1 - x)}y' + \dfrac{2}{x(1 - x)}y = 0$

(g) $y'' - \frac{(x - 3x^2)}{2(x^2 + x^3)}y' + \frac{y}{2(x^2 + x^3)} = 0$

(h) $y'' + \left(1 - \frac{1}{x}\right)y' - \frac{1}{x}y = 0$ (i) $y'' - 2y' + \left(\frac{1}{4x^2} + 1\right)y = 0$

(j) $y'' - \frac{1}{x}y' + \left(1 + \frac{1}{x^2}\right)y = 0$

7. Show that the condition $Q \le 0$ is *not* a necessary condition for the validity of Theorem 4.4.
8. For each equation of Exercise 4, find the unique solution that satisfies the initial conditions $y(0) = 0$, $y'(0) = 1$. Check your answers.
9. For the equations of Exercises 3(b), 4(b), 4(e), and 4(f), find, if possible, a solution that satisfies the boundary conditions $y(0) = y(1) = 0$. Check your answers and determine whether or not each solution is unique.
10. On the R_{11} set $x_0 = 0, x_1 = 1, x_2 = 2, \ldots, x_{10} = 10$, solve each of the following initial-value problems:

(a) $y_0 = 0, \frac{y_1 - y_0}{\Delta x} = 1, y_{n+1} = x_n + \frac{(\Delta x)y_{n-1}}{y_n}, n = 1, 2, \ldots, 9$

(b) $y_0 = 1, \frac{y_1 - y_0}{\Delta x} = 0, y_{n+1} = (\Delta x)y_n + \left(\frac{y_{n-1}}{x_n}\right)^2, n = 1, 2, \ldots, 9$

(c) $y_0 = 1, \frac{y_1 - y_0}{\Delta x} = -1, y_{n+1} = x_n \Delta x(y_n - y_{n-1}), n = 1, 2, \ldots, 9$

(d) $y_0 = 1, \frac{y_1 - y_0}{\Delta x} = 2, y_{n+1} = \Delta x(y_n - y_{n-1}), n = 1, 2, \ldots, 9$

(e) $y_0 = 1, \frac{y_1 - y_0}{\Delta x} = -1, y_{n+1} = (\Delta x)y_{n-1}, n = 1, 2, \ldots, 9$

11. On the R_5 set $x_0 = 0, x_1 = 0.1, x_2 = 0.2, x_3 = 0.3, x_4 = 0.4$, solve the following boundary-value problems:

(a) $y_{n+1} = \frac{1}{1 + \Delta x}[2y_n + (\Delta x)y_{n-1} - y_{n-1}], y(0) = 0, y(0.4) = 1$

(b) $y_{n+1} = \frac{1}{1 + \Delta x}[2y_n + (\Delta x)y_{n-1} - y_{n-1} + (\Delta x)^2],$

$y(0) = 1, y(0.4) = 0$

12. On the R_{11} set $x_n = n, n = 0, 1, \ldots, 10$, consider the boundary-value

problem

$$y_{n+1} = 2y_n - y_{n-1}, \qquad n = 1, 2, \ldots, 9,$$

$$y_0 = 0, \qquad y_{10} = 1.$$

By means of Theorem 3.1, show that the solution exists and is unique. By use of the direct method of Section 3.4, construct the solution.

13. On the R_{1001} set $x_n = n$, $n = 0, 1, 2, \ldots, 1000$, consider the boundary-value problem

$$y_{n+1} = 3y_n - 2y_{n-1}, \qquad n = 1, 2, \ldots, 999,$$

$$y_0 = 0, \qquad y_{1000} = 1.$$

By means of Theorem 3.1, show that the exact solution exists and is unique. By use of the iterative method of Section 3.5, approximate the solution.

Chapter 5

APPROXIMATE SOLUTION OF BOUNDARY-VALUE PROBLEMS FOR DIFFERENTIAL EQUATIONS

5.1. Introduction

The concepts developed thus far enable us now, in a natural way, to approximate solutions of boundary-value problems for both linear and nonlinear second-order differential equations. Throughout, for simplicity and motivation, linear problems will be explored first, and then the ideas and methods will be extended to nonlinear problems.

5.2. Approximate Solution of Linear Boundary-Value Problems

Let I be an open interval $a < x < b$ and let α and β be constants. Let $P(x)$, $Q(x)$, $R(x)$ be both bounded and continuous on I. Then we shall consider in this section approximating a solution of the boundary-value problem defined by

$$y'' + P(x)y' + Q(x)y = R(x), \tag{5.1}$$

$$y(a) = \alpha, \qquad y(b) = \beta. \tag{5.2}$$

For simplicity, it is desirable that (5.1) and (5.2) have a unique solution, and so we shall assume in addition that the sufficiency condition of Theorem 4.4 is valid—that

$$Q(x) \leq 0, \qquad a < x < b. \tag{5.3}$$

The spirit of our approach to (5.1) through (5.3) is one in which we discretize all possible aspects of the problem. In other words, we shall approximate the continuous problem by a discrete one. This will result in the algebraic problem of solving a tridiagonal system, which, when small, can be solved by the direct method of Section 3.4, and which, when large, can be solved by the iterative method of Section 3.5.

Problem (5.1) through (5.3) can be discretized by replacing the interval $a \le x \le b$, which consists of an infinite number of points, by a finite R_{n+1} point set, and by replacing the second-order differential equation (5.1) by a second-order difference equation. These will be done as follows. Divide $a \le x \le b$ into n equal parts, each of length Δx, by the R_{n+1} set $a = x_0 < x_1 < x_2 < \cdots < x_n = b$. Substitute

$$y'' \cong \frac{y_{i-1} - 2y_i + y_{i+1}}{(\Delta x)^2} \tag{5.4}$$

and any one of

$$y' \cong \frac{y_{i+1} - y_{i-1}}{2\,\Delta x}, \tag{5.5}$$

$$y' \cong \frac{y_{i+1} - y_i}{\Delta x}, \tag{5.6}$$

$$y' \cong \frac{y_i - y_{i-1}}{\Delta x} \tag{5.7}$$

into (5.1) to obtain an approximating difference equation on $I(R_{n+1})$. Thus, using (5.5), which is currently popular, one would have

$$\frac{y_{i-1} - 2y_i + y_{i+1}}{(\Delta x)^2} + P(x_i)\frac{y_{i+1} - y_{i-1}}{2\,\Delta x} + Q(x_i)y_i = R(x_i),$$
$$i = 1, 2, \ldots, n-1,$$

or, equivalently,

$$\begin{aligned} y_{i-1}[2 - P(x_i)\,\Delta x] + y_i[-4 + 2(\Delta x)^2 Q(x_i)] + y_{i+1}[2 + P(x_i)\,\Delta x] \\ = 2(\Delta x)^2 R(x_i), \qquad i = 1, 2, \ldots, n-1. \end{aligned} \tag{5.8}$$

From (5.2), it follows that if one writes down, in *order*, each of equations (5.8), as will be illustrated in the examples that follow, there results a tridiagonal linear system of $(n-1)$ equations in the $(n-1)$ unknowns $y_1, y_2, \ldots, y_{n-1}$. A solution of this system constitutes an approximate solution of the problem (5.1) through (5.3) on $I(R_{n+1})$.

Example 1. Consider the utterly simple boundary-value problem

$$y'' = 0,$$
$$y(0) = -1, \qquad y(6) = 0.$$

Divide [0,6] into six equal parts by the R_7 set $x_0 = 0$, $x_1 = 1$, $x_2 = 2$, $x_3 = 3$, $x_4 = 4$, $x_5 = 5$, $x_6 = 6$. Then (5.8) reduces to

$$y_{i-1} - 2y_i + y_{i+1} = 0, \qquad i = 1, 2, 3, 4, 5. \tag{5.9}$$

In order, for $i = 1, 2, 3, 4, 5$, this difference equation yields

$$\begin{aligned}
y_0 - 2y_1 + y_2 &= 0,\\
y_1 - 2y_2 + y_3 &= 0,\\
y_2 - 2y_3 + y_4 &= 0,\\
y_3 - 2y_4 + y_5 &= 0,\\
y_4 - 2y_5 + y_6 &= 0.
\end{aligned}$$

But, from the given boundary data, $y_0 = -1$, $y_6 = 0$, so that the above system becomes

$$\begin{aligned}
-2y_1 + y_2 &= 1,\\
y_1 - 2y_2 + y_3 &= 0,\\
y_2 - 2y_3 + y_4 &= 0,\\
y_3 - 2y_4 + y_5 &= 0,\\
y_4 - 2y_5 &= 0.
\end{aligned} \tag{5.10}$$

By Theorem 3.1, this tridiagonal system has a unique solution and by the direct method of Section 3.4, it is found to be

$$y_1 = -\tfrac{5}{6},\ y_2 = -\tfrac{2}{3},\ y_3 = -\tfrac{1}{2},\ y_4 = -\tfrac{1}{3},\ y_5 = -\tfrac{1}{6},$$

which, because of the simplicity of the given problem, coincides on $I(R_7)$ with the exact solution $y = (x/6) - 1$ of (5.1) through (5.3). It is most important to note that for both (5.9) and (5.8), the coefficient of y_i yields the main diagonal entries, those of y_{i-1} yield the entries just below the main diagonal, and those of y_{i+1} yield the entries just above the main diagonal of the matrix of the resulting tridiagonal system.

Example 2. Consider the boundary-value problem

$$y'' - y = 0, \qquad 0 < x < 3, \tag{5.11}$$

$$y(0) = 0, \qquad y(3) = 1. \tag{5.12}$$

With $\Delta x = \frac{1}{2}$, the R_7 set that divides $0 \le x \le 3$ into six equal parts is $x_0 = 0$, $x_1 = 0.5$, $x_2 = 1$, $x_3 = 1.5$, $x_4 = 2$, $x_5 = 2.5$, $x_6 = 3$. Since no approximation for y' must be made, (5.11) is approximated by the difference equation

$$\frac{y_{i-1} - 2y_i + y_{i+1}}{(\Delta x)^2} - y_i = 0, \qquad i = 1, 2, 3, 4, 5,$$

or, equivalently, by

$$y_{i-1} - \tfrac{9}{4}y_i + y_{i+1} = 0, \qquad i = 1, 2, 3, 4, 5. \tag{5.13}$$

From (5.13) it can be seen immediately that the main diagonal coefficients will be $-\frac{9}{4}$, which will make Theorem 3.1 again applicable. Using the given boundary values, the tridiagonal system that results for $i = 1, 2, 3, 4, 5$ is

$$\begin{aligned}
-\tfrac{9}{4}y_1 + y_2 \qquad\qquad\qquad &= 0,\\
y_1 - \tfrac{9}{4}y_2 + y_3 \qquad\qquad &= 0,\\
y_2 - \tfrac{9}{4}y_3 + y_4 \qquad &= 0,\\
y_3 - \tfrac{9}{4}y_4 + y_5 &= 0,\\
y_4 - \tfrac{9}{4}y_5 &= -1,
\end{aligned}$$

the solution of which, found by the direct method, is

$$y_1 = \frac{2^8 \cdot 13 \cdot 863 \cdot 5987}{3^5 \cdot 5 \cdot 7^2 \cdot 11 \cdot 13^2 \cdot 29 \cdot 101}, \qquad y_2 = \frac{2^5 \cdot 863 \cdot 5987}{3^3 \cdot 5 \cdot 7^2 \cdot 11 \cdot 13 \cdot 29 \cdot 101},$$

$$y_3 = \frac{2^4 \cdot 863 \cdot 5987}{3^5 \cdot 7^2 \cdot 11 \cdot 29 \cdot 101}, \qquad y_4 = \frac{2^2 \cdot 863 \cdot 5987}{3^3 \cdot 5 \cdot 11 \cdot 13 \cdot 29 \cdot 101},$$

$$y_5 = \frac{2^2 \cdot 29 \cdot 101}{3^2 \cdot 5 \cdot 29 \cdot 43}.$$

Considering that the given interval was divided into only six equal parts, it follows easily that on $I(R_7)$ these numerical results approximate rather well the exact values which can be obtained from the analytical solution

$$y = e^3(e^x - e^{-x})/(e^6 - 1).$$

Example 3. Consider the general linear boundary-value problem (5.1) through (5.3) and let (5.1) be approximated by (5.8). If $a \le x \le b$ is divided into n parts, it follows easily, as in the above examples, that (5.8) yields a tridiagonal linear system. It is not at all clear, however, that this system satisfies the additional conditions of Theorem 3.1. If we wish to have this assurance, we must examine (5.8) in somewhat greater detail. From (5.3) it follows that the coefficient of y_i, and hence that each main diagonal element, is negative. To be assured that the elements just above and just below the main diagonal elements are positive would require that

$$2 - P(x_i)\,\Delta x > 0,$$

$$2 + P(x_i)\,\Delta x > 0.$$

These would both be satisfied if for all x in $a \leq x \leq b$

$$2 > \Delta x |P(x)|. \tag{5.14}$$

Now since $P(x)$ has been assumed to be not only continuous on $a < x < b$ but also bounded, choose any positive upper bound M of $P(x)$ on $a < x < b$ which satisfies

$$M \geq |P(x)|, \qquad a < x < b.$$

Then the choice

$$\Delta x < \frac{2}{M} \tag{5.15}$$

would be sufficient to assure the validity of (5.14). Finally, the required diagonal dominance for (5.8) follows from (5.15), because

$$-[-4 + 2(\Delta x)^2 Q(x_i)] \geq [2 - P(x_i)\,\Delta x] + [2 + P(x_i)\,\Delta x].$$

It is interesting to observe that if one wishes to have the assurance provided by Theorem 3.1, then one can always have it merely by dividing $a \leq x \leq b$ into a sufficiently large number of parts so that (5.15) is valid.

Example 4. Again, consider general linear boundary-value problem (5.1) through (5.3). In Example 3 we saw that if one approximates (5.1) by (5.8), and if one wishes to satisfy the conditions of Theorem 3.1, then the number of parts into which one can divide $a \leq x \leq b$ is no longer a free choice but must be governed by some condition comparable to (5.15). Let us show how (5.1) through (5.3) can be approached somewhat differently so that one is still free to divide $[a,b]$ into any number of parts that he may wish and still retain the applicability of Theorem 3.1. Suppose that (5.1) is approximated at each point x_i, $i = 1, 2, \ldots, n - 1$, of a given R_{n+1} set as follows. Approximate y'' by (5.4). To approximate y', first examine $P(x_i)$. If $P(x_i) \geq 0$, use approximation (5.6), while if $P(x_i) < 0$, use approximation (5.7). Thus (5.1) is approximated as follows. If $P(x_i) \geq 0$, use

$$\frac{y_{i-1} - 2y_i + y_{i+1}}{(\Delta x)^2} + P(x_i)\frac{y_{i+1} - y_i}{\Delta x} + Q(x_i)y_i = R(x_i)$$

or, in equivalent form,

$$y_{i-1} + y_i[-2 - P(x_i)\,\Delta x + Q(x_i)(\Delta x)^2] + y_{i+1}[1 + P(x_i)\,\Delta x] = (\Delta x)^2 R(x_i),$$

whereas if $P(x_i) < 0$, use

$$\frac{y_{i-1} - 2y_i + y_{i+1}}{(\Delta x)^2} + P(x_i)\frac{y_i - y_{i-1}}{\Delta x} + Q(x_i)y_i = R(x_i),$$

or, in equivalent form,

$$y_{i-1}[1 - P(x_i)\,\Delta x] + y_i[-2 + P(x_i)\,\Delta x + Q(x_i)(\Delta x)^2] + y_{i+1} = (\Delta x)^2 R(x_i).$$

Such an approach always maintains the diagonal dominance of the resulting system, for recall that it is always the coefficient of y_i which determines the main diagonal element; the coefficients of y_{i-1} and y_{i+1} determine the elements just above and just below the main diagonal. In each of the above cases the coefficient of y_i retains its negativeness for all Δx and, indeed, all the conditions of Theorem 3.1 are valid for all Δx.

With regard to Examples 1 through 4, it should be noted that successive overrelaxation is known to converge for all initial guesses and for all ω in the range $0 < \omega < 2$ (see, e.g., Varga). Although nonconstructive results are available for determining the optimum ω for the iteration, in practice one often selects a set of ω's in the range $0 < \omega < 2$, lets each run for, say, 10 or 15 iterations with a zero initial vector, and then chooses that ω which seems to be giving the most rapid convergence. Studies in which ω is allowed to vary have been initiated only recently.

5.3. Approximate Solution of Nonlinear Boundary-Value Problems

The ideas of Section 5.2 for linear problems extend directly to the following important class of nonlinear boundary-value problems. For a, b, α, β constants, with $a < b$, consider the general boundary-value problem

(5.16) $$y'' = f(x,y,y'),$$

(5.17) $$y(a) = \alpha, \qquad y(b) = \beta,$$

where $f(x,y,z)$ has continuous first-order partial derivatives for all y, for all z, and for x in the range $a \le x \le b$. Then we shall study that special subclass of (5.16) and (5.17) for which the following is also valid. For all y, all z, and all x in $a \leq x \leq b$, assume that

$$(5.18) \qquad \frac{\partial f(x,y,z)}{\partial y} \geq 0, \qquad \left|\frac{\partial f(x,y,z)}{\partial z}\right| \leq A,$$

where A is a positive constant.

Boundary-value problem (5.16) through (5.18) has a unique solution (see, e.g., Keller). Moreover, when (5.16) has the linear form (5.1), the first condition of (5.18) is exactly that $Q(x) \leq 0$.

Problem (5.16) through (5.18) is discretized by replacing the interval $a \leq x \leq b$ by an R_{n+1} set, as in Section 5.2, and by approximating differential equation (5.16) by a difference equation that uses (5.4) and (5.5), that is, by

$$(5.19) \qquad \frac{y_{i-1} - 2y_i + y_{i+1}}{(\Delta x)^2} = f\left(x_i, y_i, \frac{y_{i+1} - y_{i-1}}{2\,\Delta x}\right), \qquad i = 1, 2, \ldots, n-1.$$

One writes down, in order, each of (5.19) for $i = 1, 2, \ldots, n-1$ and, by (5.17), this results in a system of nonlinear equations in $y_1, y_2, \ldots, y_{n-1}$. A solution of this system constitutes an approximate solution of (5.16) through (5.18) on $I(R_{n+1})$.

Example. For the one-dimensional radiation equation

$$(5.20) \qquad y'' = e^y,$$

consider the boundary-value problem with $y(0) = 0$, $y(1) = 0$. The problem has a unique solution and we proceed to approximate it numerically as follows. Divide [0,1] into three equal parts by the R_4 set $x_0 = 0$, $x_1 = \frac{1}{3}$, $x_2 = \frac{2}{3}$, $x_3 = 1$, and approximate (5.17) by

$$\frac{y_{i-1} - 2y_i + y_{i+1}}{(\frac{1}{3})^2} = e^{y_i}, \qquad i = 1, 2.$$

Thus

$$9y_0 - 18y_1 + 9y_2 = e^{y_1},$$
$$9y_1 - 18y_2 + 9y_3 = e^{y_2},$$

so that from the given boundary conditions, one has

$$e^{y_1} + 18y_1 - 9y_2 = 0,$$
$$e^{y_2} + 18y_2 - 9y_1 = 0.$$

This system is completely analogous to that solved in (3.52) through (3.55).

In this case the iteration takes the form

$$y_1^{(k+1)} = y_1^{(k)} - \omega \frac{e^{y_1^{(k)}} + 18y_1^{(k)} - 9y_2^{(k)}}{e^{y_1^{(k)}} + 18},$$

$$y_2^{(k+1)} = y_2^{(k)} - \omega \frac{e^{y_2^{(k)}} + 18y_2^{(k)} - 9y_1^{(k+1)}}{e^{y_2^{(k)}} + 18},$$

and must be calculated on a high-speed computer to obtain a high degree of accuracy.

It should be noted that it is known (Bers) that for all sufficiently small Δx, the nonlinear system generated by the numerical method of this section has a unique solution. However, the methods required for the proof of this result are of an advanced topological nature and beyond the scope of this book. That the resulting equations can always be solved by the generalized Newton's method for ω in a proper subset of the range $0 < \omega < 2$ was proved by Schechter.

5.4. Convergence

The methods of Sections 5.2 and 5.3 will have a relatively firm mathematical foundation if we can show finally that as Δx goes to zero, the numerical solution converges to the analytical solution of a given boundary-value problem. For this purpose, consider first the special linear boundary-value problem

$$y'' + P(x)y' + Q(x)y = R(x), \tag{5.21}$$

$$y(a) = \alpha, \qquad y(b) = \beta, \tag{5.22}$$

as posed in Section 5.2, but with (5.3) replaced by the stronger condition that for some $\varepsilon > 0$,

$$Q(x) < -\varepsilon. \tag{5.23}$$

We shall assume that (5.21) through (5.23) has been solved by the numerical method of Section 5.2 with the approximations (5.4) and (5.5) substituted into (5.21) to yield a difference-equation approximation. For any function $v(x,y)$ which is continuous on $a \le x \le b$ and which has continuous derivatives on $a < x < b$, the operators L and $L_{\Delta x}$ are defined at each point x_i, $i = 1, 2, \ldots, n-1$ of the R_{n+1} set which subdivides $a \le x \le b$ by

$$Lv(x_i) = v''(x_i) + P(x_i)v'(x_i) + Q(x_i)v_i, \tag{5.24}$$

$$L_{\Delta x}v(x_i) = \frac{v_{i-1} - 2v_i + v_{i+1}}{(\Delta x)^2} + P(x_i)\frac{v_{i+1} - v_{i-1}}{2\,\Delta x} + Q(x_i)v_i. \tag{5.25}$$

Finally, by (5.23), and because $P(x)$ and $Q(x)$ in Section 5.2 have been assumed to be not only continuous but also bounded on $a < x < b$, let positive bounds M, N_1, N_2 be defined by

$$M \geq |P(x)|, \qquad a < x < b, \tag{5.26}$$

$$N_1 \geq -Q(x) \geq N_2 > 0, \qquad a < x < b. \tag{5.27}$$

Lemma 5.1. Let v be a discrete function defined on the R_{n+1} set which subdivides $a \leq x \leq b$ into n equal parts. Let Δx satisfy (5.15), that is, $\Delta x < 2/M$, where M is defined by (5.26). Define C by

$$C = \max\left[1, \frac{1}{N_2}\right].$$

Then

$$|v(x_i)| \leq C\{\max[|v_0|, |v_n|] + \max_{1 \leq j \leq n-1} |L_{\Delta x}v(x_j)|\}, \qquad i = 0, 1, 2, \ldots, n. \tag{5.28}$$

Proof. If $\max|v(x_i)|$ occurs for $i = 0$ or $i = n$, then (5.28) is valid, since $C \geq 1$. *We proceed then under the assumption that* $\max|v(x_i)|$ *is attained for i equal to one of* $1, 2, 3, \ldots, n-1$.

Suppose, for definiteness, that the system of algebraic equations in our numerical method is written in the particular form

$$\frac{(\Delta x)^2}{2} L_{\Delta x}v(x_i) = \frac{(\Delta x)^2}{2} R(x_i), \qquad i = 1, 2, \ldots, n-1,$$

so that, from (5.25),

$$\frac{(\Delta x)^2}{2} L_{\Delta x}v(x_i) = v_{i-1}\left[\frac{1}{2} - \frac{\Delta x}{4}P(x_i)\right] + v_i\left[-1 + \frac{(\Delta x)^2}{2}Q(x_i)\right] + v_{i+1}\left[\frac{1}{2} + \frac{\Delta x}{4}P(x_i)\right], \qquad i = 1, 2, \ldots, n-1. \tag{5.29}$$

Thus the coefficient matrix of the resulting tridiagonal system will have main diagonal elements

$$a_{ii} = -1 + \frac{(\Delta x)^2}{2}Q(x_i) < -1, \tag{5.30}$$

will have elements below the main diagonal

$$a_{i,i-1} = \frac{1}{2} - \frac{\Delta x}{4}P(x_i) > 0, \qquad i = 2, 3, \ldots, n, \tag{5.31}$$

and will have elements above the main diagonal

$$a_{i,i+1} = \frac{1}{2} + \frac{\Delta x}{4} P(x_i) > 0, \qquad i = 1, 2, \ldots, n-1. \tag{5.32}$$

Consider now (5.29) in the equivalent form

$$-v_i\left[1 - \frac{(\Delta x)^2}{2} Q(x_i)\right] = -v_{i-1}\left[\frac{1}{2} - \frac{\Delta x}{4} P(x_i)\right] - v_{i+1}\left[\frac{1}{2} + \frac{\Delta x}{4} P(x_i)\right] + \frac{(\Delta x)^2}{2} L_{\Delta x} v(x_i), \qquad i = 1, 2, \ldots, n-1.$$

Taking absolute values in this last relation implies readily, with the aid of (5.27) and (5.30) through (5.32), that

$$\left[1 + \frac{(\Delta x)^2}{2} N_2\right] \cdot |v_i| \le \max_{j=1,2,\ldots,n-1} |v_j| + \frac{(\Delta x)^2}{2} \max_{j=1,2,\ldots,n-1} |L_{\Delta x} v(x_j)|, \qquad i = 1, 2, \ldots, n-1.$$

Since the right-hand side of this inequality is independent of i, it follows that if one chooses that value of i for which $|v_i|$ is a maximum, then the inequality yields

$$N_2 \max_i |v_i| \le \max_i |L_{\Delta x} v(x_i)|, \qquad i = 1, 2, \ldots, n-1.$$

Hence

$$\max_i |v_i| \le C \max_i |L_{\Delta x} v(x_i)| \le C\{\max[|v_0|, |v_n|] + \max_i |L_{\Delta x} v(x_i)|\}, \qquad i = 1, 2, \ldots, n-1,$$

and the lemma is proved.

With the aid of Lemma 5.1, we can now prove a major theorem related to the question of convergence.

Theorem 5.1. Let $y(x)$ be the solution of boundary-value problem (5.21) through (5.23) and let u_i, $i = 1, 2, \ldots, n-1$ be the numerical solution on $I(R_{n+1})$ generated by the numerical method of Section 5.3. In the notation of this section and under the assumptions of Lemma 5.1, if $y(x)$ has continuous derivatives up to and including order 4 on $a \le x \le b$, then

$$|u_i - y(x_i)| \le C \frac{(\Delta x)^2}{12} (M_4 + 2MM_3), \qquad i = 0, 1, 2, \ldots, n, \tag{5.33}$$

where

$$M_r = \max_{a \le x \le b} \left|\frac{d^r y}{dx^r}\right|, \qquad r = 3, 4. \tag{5.34}$$

Proof. The validity of (5.33) for $i = 0, i = n$ is a direct consequence of the numerical method, since $u_0 = y(x_0), u_n = y(x_n)$. Hence we need only consider $i = 1, 2, \ldots, n - 1$.

Set

$$v_i = u_i - y(x_i)$$

into (5.28). Since $v_0 = v_n = 0$, one has

$$(5.35) \qquad |u_i - y(x_i)| \leq C \max_i |L_{\Delta x}[u_i - y(x_i)]|, \qquad i = 1, 2, \ldots, n - 1.$$

But

$$\begin{aligned} L_{\Delta x}[u_i - y(x_i)] &= L_{\Delta x}u_i - L_{\Delta x}y(x_i) + Ly(x_i) - Ly(x_i) \\ &= R(x_i) - L_{\Delta x}y(x_i) + Ly(x_i) - R(x_i) \\ &= y''(x_i) + P(x_i)y'(x_i) + Q(x_i)y_i \\ &\quad - \frac{y_{i-1} - 2y_i + y_{i+1}}{(\Delta x)^2} - P(x_i)\frac{y_{i+1} - y_{i-1}}{2\,\Delta x} - Q(x_i)y_i, \end{aligned}$$

which, by insertion of Taylor expansions through fourth order for y_{i-1} and y_{i+1} into $(y_{i-1} - 2y_i + y_{i+1})/(\Delta x)^2$ and through third order for y_{i-1} and y_{i+1} into $(y_{i+1} - y_{i-1})/(2\Delta x)$ implies that

$$\begin{aligned} L_{\Delta x}[u_i - y(x_i)] = &-\frac{(\Delta x)^2}{24}\frac{d^4y(\xi_1)}{dx^4} - \frac{(\Delta x)^2}{24}\frac{d^4y(\xi_2)}{dx^4} \\ &- \frac{P(x_i)(\Delta x)^2}{12}\frac{d^3y(\xi_3)}{dx^3} - \frac{P(x_i)(\Delta x)^2}{12}\frac{d^3y(\xi_4)}{dx^3}. \end{aligned}$$

Thus

$$|L_{\Delta x}[u_i - y(x_i)]| \leq \frac{(\Delta x)^2}{12}M_4 + \frac{(\Delta x)^2}{6}MM_3,$$

which, upon substitution into (5.35) yields (5.33), and the theorem is proved.

The inequality (5.33) is called an *error estimate*. Although it is nonconstructive in the sense that we have no visible means of estimating M_3 and M_4, it does say that, for the class of problems to which the theorem is applicable, one has $u_i \to y(x_i)$ as $\Delta x \to 0$, which is what we mean by *convergence* of the numerical solution to the analytical solution.

Extension of Theorem 5.1 to more general classes of both linear and nonlinear problems can be accomplished in the same spirit as that of Theorem 5.1, but with *much* more effort required [see, e.g., Bers; Greenspan (3)]. Thus convergence theorems are available for the linear and nonlinear problems of this chapter, but because of the extensive mathematical analysis required for their proofs, they must be left to the reading of the more ambitious and mathematically inclined student.

5.5. Remarks

Of the large variety of other methods available for the approximate solution of boundary-value problems, some of the more valuable ones include variational methods [Collatz (1)], continuous methods [Fox (2)], integral equation methods [Keller; Rall (2)], collocation [Collatz (1)], and shooting methods (Keller).

EXERCISES

1. To how many decimal places do the approximate and exact solutions given for (5.11) and (5.12) agree?
2. Show that each of the following boundary-value problems has a unique solution by applying Theorem 4.4. Solve each, first exactly, then numerically with $x = \frac{1}{5}$:

   (a) $y'' = 0$, $y(0) = 0$, $y(1) = 1$
   (b) $y'' = x$, $y(0) = 0$, $y(1) = 1$
   (c) $y'' - y = 0$, $y(0) = 0$, $y(1) = 1$
   (d) $y'' - 4y' - 3y = 0$, $y(0) = 0$, $y(1) = 1$
   (e) $y'' + 4y' - 4y = 0$, $y(0) = 0$, $y(1) = 1$
   (f) $y'' - y = x$, $y(0) = 0$, $y(1) = 1$
3. Find an approximate solution for each of the following boundary-value problems. Be sure that the tridiagonal system which you set up has a unique solution.
   (a) $y'' - y' - 2y = 0$, $y(0) = 0$, $y(1) = 2$
   (b) $y'' + xy' - y = x$, $y(0) = -1$, $y(1) = 1$
   (c) $y'' + (\sin x)y' - y = 0$, $y(0) = 0$, $y(2\pi) = -1$
4. Find an approximate solution for each of the following boundary-value problems:
   (a) $y'' = e^y$, $y(0) = 0$, $y(1) = 1$ (b) $y'' + y' = e^y$, $y(0) = 0$, $y(1) = 1$
   (c) $y'' = y^2$, $y(0) = 1$, $y(1) = 1$ (d) $y'' = y^3$, $y(0) = 0$, $y(1) = 1$
   (e) $y'' = y^5$, $y(0) = 0$, $y(1) = 1$
5. For the following choices of $v(x)$, $P(x)$, and $Q(x)$, determine $Lv(x_i)$ from (5.24) and $L_{\Delta x}v(x_i)$ from (5.25):
   (a) $v(x) = 1$, $P(x) = 1$, $Q(x) = -1$
   (b) $v(x) = x$, $P(x) = 1$, $Q(x) = -1$
   (c) $v(x) = x^2$, $P(x) = -1$, $Q(x) = -1$

(d) $v(x) = x^3$, $P(x) = -1$, $Q(x) = 0$
(e) $v(x) = e^x$, $P(x) = x$, $Q(x) = -x^2$
(f) $v(x) = 1 + x^2$, $P(x) = 0$, $Q(x) = -1$

6. For each of the following boundary-value problems, estimate the constant C of Lemma 5.1:
(a) $y'' + y' - y = 0$, $y(0) = y(1) = 1$
(b) $y'' - 2y' - 3y = 0$, $y(0) = y(1) = 1$
(c) $y'' + 3y' - (1 + x)y = 0$, $y(0) = y(1) = 1$
(d) $y'' - 2y' - (e^x)y = 0$, $y(0) = y(1) = 1$

Chapter 6

APPROXIMATE SOLUTION OF INITIAL-VALUE PROBLEMS FOR ORDINARY DIFFERENTIAL EQUATIONS

6.1. Introduction

Besides boundary-value problems, initial-value problems are also of basic importance in the study and application of second-order ordinary differential equations. But, unlike boundary-value problems, initial-value problems are usually approached by transforming them into *systems* of first-order equations as follows. The initial-value problem

$$y'' = f(x,y,y'), \tag{6.1}$$

$$y(a) = \alpha, \qquad y'(a) = \beta \tag{6.2}$$

is equivalent, under the change of variables

$$y' = v, \tag{6.3}$$

to the system

$$y' = v, \qquad y(a) = \alpha, \tag{6.4}$$

$$v' = f(x,y,v), \qquad v(a) = \beta \tag{6.5}$$

of two first-order differential equations in the two unknown functions y and v, each of which is given initially at $x = a$.

Example. Let I be the interval defined by $x > 0$. Then on I the initial-value problem

$$y'' = y'e^x + \sin y,$$

$$y(0) = \frac{\pi}{4}, \qquad y'(0) = 0$$

is equivalent to the system

$$y' = v, \qquad y(0) = \frac{\pi}{4},$$

$$v' = ve^x + \sin y, \qquad v(0) = 0.$$

Because of its simplicity and convenience, the above approach will also be followed here. But we shall begin in a seemingly less ambitious fashion by considering only one first-order equation in one unknown function, for the important reason that *all* the mathematical methods and theory developed for such an equation will extend with exceptional ease to systems of the form (6.4) and (6.5). It must be emphasized, however, that only systems of at least two such equations have significant physical applications.

We shall assume throughout that any generally stated initial-value problem has a unique solution. For a large variety of sufficient conditions which assure such existence and uniqueness, consult Coddington and Levinson.

6.2. First-Order Differential Equations

Let I be an open interval of real numbers. For x in I and for all real numbers u, let $f(x,u)$ be defined and continuous. Then

$$y' = f(x,y) \tag{6.6}$$

is said to be a first-order differential equation on I. A function $y = g(x)$ is said to be a solution of (6.6) on I if and only if

(a) $y = g(x)$ is defined and continuous on $\bar{I}$;
(b) $y = g(x)$ has continuous first-order derivatives on I; and
(c) $g'(x) = f(x,g(x))$ is an *identity* in x on I.

Example 1. On $0 < x < 1$,

$$y' = xy$$

is a first-order differential equation, solutions of which are $y_1 = e^{x^2/2}$, $y_2 = 2e^{x^2/2}$, and $y_3 = -5e^{x^2/2}$.

Example 2. On $0 < x$,

$$y' = xy^2 - x^5 + 2x$$

is a first-order differential equation, a solution of which is $y = x^2$.

If I is a given open interval $a < x$ or $a < x < b$, and if α is a given real number, then the problem of constructing a solution on I of (6.6) which in addition satisfies

(6.7) $$y(a) = \alpha$$

is called an initial-value problem.

Example 3. The problem of finding a solution of

$$y' = x - y + y^3$$

on $x > 0$ which satisfies

$$y(0) = \pi$$

is an initial-value problem.

General analytical methods for solving a first-order differential equation or an initial-value problem for such an equation are restricted largely to the class of *linear equations*, which are defined as follows. Equation (6.6) is said to be linear on I if $f(x,y)$ is linear in y, that is, if (6.6) is of the special form

(6.8) $$y' = P(x)y + Q(x),$$

where P and Q are continuous on I. The well-known methods of multiplying both sides of (6.8) by an integrating factor of the form $\exp[-\int P(x)\,dx]$, where $\int P(x)\,dx$ is any particular antiderivative of $P(x)$, can always be used to solve (6.8), as is illustrated in the following examples.

Example 4. On $x > 0$, consider

(6.9) $$y' = xy + e^{x^2/2}.$$

Rewrite (6.9) in the equivalent form

(6.10) $$y' - xy = e^{x^2/2}.$$

Multiply both sides of (6.10) by the integrating factor $e^{-x^2/2}$ to yield

$$y' \cdot e^{-x^2/2} - xy \cdot e^{-x^2/2} = 1,$$

which, by the rule for the derivative of a product, contracts to

(6.11) $$\frac{d}{dx}(y \cdot e^{-x^2/2}) = 1.$$

Integrating both sides of (6.11) then implies that

(6.12) $$y \cdot e^{-x^2/2} = x + C,$$

where C is an arbitrary constant. Finally, from (6.12) one has a one-parameter family of solutions of (6.9) of the form

$$y = xe^{x^2/2} + Ce^{x^2/2}. \tag{6.13}$$

Example 5. On $x > 0$, the unique solution of the initial-value problem

$$y' = xy + e^{x^2/2}, \tag{6.14}$$

$$y(0) = -1 \tag{6.15}$$

is found from (6.13) by setting $x = 0$, $y = -1$ and determining that C must have the value -1. Thus the solution is

$$y = xe^{x^2/2} - e^{x^2/2}.$$

When (6.6) is not linear, initial-value problem (6.6) and (6.7) can rarely be solved analytically and one must consider a numerical method. Nevertheless, a casual glance at a compendium such as that of Kamke, in which a number of highly specialized, analytical techniques have been catalogued, may, on rare occasions, prove to be of value.

6.3. First-Order Difference Equations

In analogy with first-order differential equations, concepts related to first-order difference equations can be defined and developed as follows.

Let $x_0, x_1, \ldots, x_n$ be an R_{n+1} set. As defined previously, the interior of R_{n+1}, denoted by $I(R_{n+1})$, is the set $x_1, x_2, \ldots, x_{n-1}$. Let $I^0(R_{n+1})$ be the set $I(R_{n+1})$ plus the point x_0. Then an equation of the form

$$y_{i+1} = F(x_i, y_i, \Delta x), \qquad i = 0, 1, 2, \ldots, n-1, \tag{6.16}$$

where, in fact, F need not depend on x_i or Δx but must depend on y_i, is called a first-order difference equation on $I^0(R_{n+1})$.

Example 1. Let R_5 be the set $x_0 = 0$, $x_1 = 1$, $x_2 = 2$, $x_3 = 3$, $x_4 = 4$. On $I^0(R_5)$ each of the following is a first-order difference equation:

$$y_{i+1} = y_i + 2(\Delta x)x_i, \qquad i = 0, 1, 2, 3,$$

$$y_{i+1} = y_i(x_i)^2, \qquad i = 0, 1, 2, 3,$$

$$y_{i+1} = (\Delta x)e^{y_i}, \qquad i = 0, 1, 2, 3.$$

On the R_{n+1} set $x_0, x_1, \ldots, x_n$, the problem of finding from a prescribed value $y_0 = \alpha$ values $y_0, y_1, \ldots, y_n$ which satisfy (6.16) for $i = 0, 1, 2, \ldots, n-1$, is called an initial-value problem for (6.16). Any set $y_0, y_1, \ldots, y_n$ with

$y_0 = \alpha$ which satisfies the given difference equation $I^0(R_{n+1})$ is called a solution of the problem.

Example 2. The problem of finding values y_0, y_1, y_2, y_3, y_4 on the R_5 set $x_0 = 0, x_1 = 1, x_2 = 2, x_3 = 3, x_4 = 4$ such that

$$y_0 = 1 \tag{6.17}$$

and

$$y_{i+1} = (y_i)^2 - (x_i)^2, \qquad i = 0, 1, 2, 3, \tag{6.18}$$

is an initial-value problem. From (6.17), it follows in a recursive fashion from (6.18) that

$$y_1 = (y_0)^2 - (x_0)^2 = 1,$$
$$y_2 = (y_1)^2 - (x_1)^2 = 0,$$
$$y_3 = (y_2)^2 - (x_2)^2 = -4,$$
$$y_4 = (y_3)^2 - (x_3)^2 = 7,$$

so that a solution of initial-value problem (6.17) and (6.18) is $y_0 = 1$, $y_1 = 1, y_2 = 0, y_3 = -4, y_4 = -7$. The existence and uniqueness of this solution follows directly from the following theorem.

Theorem 6.1. If F in (6.16) is always defined, then the solution of initial value problem (6.16) and (6.17) always exists and is unique.

Proof. Since y_0 is known, the validity of the theorem is a direct consequence of the recursive nature of (6.16).

6.4. Euler's Method

The following simple numerical method for approximating a solution of initial-value problem (6.6) and (6.7) was devised by Euler in 1768.

If I is of the form $a < x < b$, divide the interval into n equal parts, each of length Δx, by the R_{n+1} set $a = x_0 < x_1 < x_2 < \cdots < x_n = b$. If I is of the form $a < x$, choose any value $b > a$ and then subdivide $a < x < b$ in the fashion just described. Next approximate y' at each point of $I^0(R_{n+1})$ by the forward-difference approximation,

$$y' \cong \frac{y_{i+1} - y_i}{\Delta x}, \tag{6.19}$$

and on $I^0(R_{n+1})$ approximate (6.6) by

$$\frac{y_{i+1} - y_i}{\Delta x} = f(x_i, y_i)$$

or, equivalently, by

$$y_{i+1} = y_i + (\Delta x) f(x_i, y_i). \tag{6.20}$$

Beginning with

$$y_0 = \alpha \tag{6.21}$$

one now solves (6.20) and (6.21) in the fashion indicated in Section 6.3 and takes the resulting discrete function to be an approximate solution of (6.6) and (6.7) on the finite set of points R_{n+1}.

Example. Consider the continuous initial-value problem

$$y' = -y + x, \qquad 0 < x < 1, \tag{6.22}$$

$$y(0) = 1. \tag{6.23}$$

For $\Delta x = 0.1$, Euler's method replaces (6.22) and (6.23) with the discrete initial-value problem

$$y_{i+1} = (0.9)y_i + (0.1)x_i, \qquad i = 0, 1, 2, \ldots, 9 \tag{6.24}$$

$$y_0 = 1, \tag{6.25}$$

where $x_0 = 0$, $x_1 = 0.1$, $x_2 = 0.2$, $x_3 = 0.3$, $x_4 = 0.4$, $x_5 = 0.5$, $x_6 = 0.6$, $x_7 = 0.7$, $x_8 = 0.8$, $x_9 = 0.9$, $x_{10} = 1.0$. Then, as shown in Section 6.3, the solution on R_{11} of (6.24) and (6.25) is

$$\begin{aligned} &y_0 = 1, && y_1 = 0.900, && y_2 = 0.820, && y_3 = 0.758, \\ &y_4 = 0.712, && y_5 = 0.681, && y_6 = 0.663, && y_7 = 0.657, \\ &y_8 = 0.661, && y_9 = 0.675, && y_{10} = 0.697. && \end{aligned} \tag{6.26}$$

The resulting discrete function constitutes an approximate solution on R_{11} of the exact solution of (6.22) and (6.23). It is of interest to note, however, that (6.22) is linear, so that, as shown in Section 5.2, one can construct the exact solution of (6.22) and (6.23), which is

$$y = x - 1 + 2e^{-x}. \tag{6.27}$$

From (6.27) one has

$$y(0) = 1, \quad y(0.1) = 0.910, \quad y(0.2) = 0.837,$$
$$y(0.3) = 0.782, \quad y(0.4) = 0.741, \quad y(0.5) = 0.713,$$
$$y(0.6) = 0.698, \quad y(0.7) = 0.693, \quad y(0.8) = 0.699, \tag{6.28}$$
$$y(0.9) = 0.713, \quad y(1.0) = 0.736.$$

6.5. Convergence

In order to study the error in the numerical solution generated by Euler's method, let us consider I to be an interval of the form $a < x < b$ and let us assume that $f(x,y)$ is continuous and has continuous and bounded first-order partial derivatives for $a \leq x \leq b$ and for all real values y. If y_i denotes the analytical solution of (6.6) and (6.7) at point x_i, while $\bar{y}_i$ denotes the numerical solution at x_i generated by (6.20) and (6.21), then the error $|E_i|$ at point x_i is defined by

$$|E_i| = |y_i - \bar{y}_i|, \qquad i = 0, 1, 2, \ldots, n. \tag{6.29}$$

The study of $|E_i|$ will be facilitated by the following lemma.

Lemma 6.1. If the numbers $|E_i|$, $i = 0, 1, \ldots, n$, satisfy

$$|E_{i+1}| \leq A|E_i| + B, \qquad i = 0, 1, \ldots, n-1, \tag{6.30}$$

where A and B are nonnegative constants, then, for $A = 1$,

$$|E_i| \leq |E_0| + iB, \qquad i = 1, 2, \ldots, n \tag{6.31}$$

while, for $A \neq 1$,

$$|E_i| \leq A^i|E_0| + \frac{A^i - 1}{A - 1}B, \qquad i = 1, 2, \ldots, n. \tag{6.32}$$

Proof. Suppose first that $A = 1$. Then from (6.30), for $i = 0$,

$$|E_1| \leq |E_0| + B. \tag{6.33}$$

From (6.30) with $i = 1$, one has

$$|E_2| \leq |E_1| + B,$$

which, from (6.33), implies that

$$|E_2| \leq |E_0| + 2B. \tag{6.34}$$

From (6.30) with $i = 2$, one has

$$|E_3| \leq |E_2| + B,$$

which, from (6.34), implies that

$$|E_3| \leq |E_0| + 3B. \tag{6.35}$$

In the indicated fashion, (6.31) follows readily by induction.

Suppose next that $A \neq 1$. Then, for $i = 0$, (6.30) yields

$$|E_1| \leq A|E_0| + B, \tag{6.36}$$

and for $i = 1$,

$$|E_2| \leq A|E_1| + B.$$

Substitution of (6.36) into the latter inequality yields

$$|E_2| \leq A(A|E_0| + B) + B$$

or

$$|E_2| \leq A^2|E_0| + (A + 1)B. \tag{6.37}$$

For $i = 2$, (6.30) implies that

$$|E_3| \leq A|E_2| + B,$$

which, from (6.37), implies that

$$|E_3| \leq A(A^2|E_0| + (A + 1)B) + B$$

or

$$|E_3| \leq A^3|E_0| + (A^2 + A + 1)B.$$

Similarly, it follows that

$$|E_4| \leq A^4|E_0| + (A^3 + A^2 + A + 1)B$$

and, by induction, that

$$|E_i| \leq A^i|E_0| + (A^{i-1} + A^{i-2} + A^{i-3} + \cdots + A + 1)B. \tag{6.38}$$

But

$$1 + A + A^2 + \cdots + A^{i-2} + A^{i-1}$$

is the sum of the geometric progression $1, A, A^2, \ldots, A^{i-2}, A^{i-1}$, whose first term is 1 and whose ratio is A, from which (6.32) follows, and the lemma is proved.

Theorem 6.2. Let I be the interval $a < x < b$. Let $y(x)$ be the solution of (6.6) and (6.7) on I and let $\bar{y}_i$, $i = 0, 1, \ldots, n$ be the numerical solution generated by (6.20) and (6.21). Assume that $y'(x)$ and $y''(x)$ exist and are continuous on $a \le x \le b$, and that $\partial f/\partial y$ exists, is continuous, and is bounded for $a \le x \le b$ and for all y. Hence there exist positive constants M and N such that

$$|y''(x)| \le N, \qquad a \le x \le b, \tag{6.39}$$

$$\left|\frac{\partial f}{\partial y}\right| \le M, \qquad a \le x \le b, \quad -\infty < y < \infty. \tag{6.40}$$

Then

$$|E_i| \le \frac{\{[1 + M\,\Delta x]^i - 1\}N\,\Delta x}{2M}, \qquad i = 0, 1, 2, \ldots, n. \tag{6.41}$$

Proof. By use of Taylor expansions and the mean value theorem, one has that

$$\begin{aligned}
|E_{i+1}| &= |y_{i+1} - \bar{y}_{i+1}| \\
&= \left|\left[y_i + (\Delta x)y_i' + \frac{(\Delta x)^2}{2}y''(\xi_i)\right] - [\bar{y}_i + (\Delta x)f(x_i, \bar{y}_i)]\right| \\
&= \left|(y_i - \bar{y}_i) + \Delta x[y_i' - f(x_i,\bar{y}_i)] + \frac{(\Delta x)^2}{2}y''(\xi_i)\right| \\
&= \left|(y_i - \bar{y}_i) + \Delta x[f(x_i,y_i) - f(x_i,\bar{y}_i)] + \frac{(\Delta x)^2}{2}y''(\xi_i)\right| \\
&= \left|(y_i - \bar{y}_i) + (\Delta x)f_y(x_i,\eta_i)(y_i - \bar{y}_i) + \frac{(\Delta x)^2}{2}y''(\xi_i)\right| \\
&= \left|(y_i - \bar{y}_i)[1 + (\Delta x)f_y] + \frac{(\Delta x)^2}{2}y''\right| \\
&\le |y_i - \bar{y}_i| \cdot [1 + (\Delta x)M] + \frac{(\Delta x)^2}{2}N \\
&= |E_i|[1 + M\,\Delta x] + \frac{(\Delta x)^2}{2}N.
\end{aligned}$$

Thus

$$|E_{i+1}| \le [1 + M\,\Delta x]|E_i| + \frac{(\Delta x)^2}{2}N, \qquad i = 0, 1, \ldots, n - 1.$$

Since $E_0 = |y_0 - \bar{y}_0| = |\alpha - \alpha| = 0$, it follows from Lemma 5.1 that

$$|E_i| \le \frac{[1 + M\,\Delta x]^i - 1}{M\,\Delta x}\,\frac{(\Delta x)^2}{2}N = \frac{\{[1 + M\,\Delta x]^i - 1\}N\,\Delta x}{2M},$$

and the theorem is proved.

Error estimate (6.41) is in many ways impractical. Not only does it often turn out to be too large to be of value, but it in no way includes the phenomenon of computer roundoff error, which is always present in actual computations. If one neglects the roundoff problem at present, then (6.41) is useful to show that the numerical solution converges to the analytical solution as $\Delta x \to 0$. This will be done by means of Theorem 6.3. Problems related to roundoff will be studied in greater detail in Section 6.10 and in Chapter 8.

Theorem 6.3. Under the assumptions of Theorem 6.2, it is valid that

$$\lim_{\Delta x \to 0} |E_i| = 0.$$

Proof. From (6.41),

$$\begin{aligned} |E_i| &\le \frac{\{[1 + M\,\Delta x]^n - 1\}N\,\Delta x}{2M} \\ &= \frac{\{[1 + M\,\Delta x]^{(b-a)/\Delta x} - 1\}N\,\Delta x}{2M} \\ &= \frac{\{[1 + M\,\Delta x]^{[1/(M\,\Delta x)]M(b-a)} - 1\}N\,\Delta x}{2M} \\ &= \frac{\{([1 + M\,\Delta x]^{1/(M\,\Delta x)})^{M(b-a)} - 1\}N\,\Delta x}{2M}. \end{aligned}$$

But, since

$$\lim_{\Delta x \to 0} ([1 + M\,\Delta x]^{1/(M\,\Delta x)})^{M(b-a)} = e^{M(b-a)},$$

it follows that

$$\begin{aligned} \lim_{\Delta x \to 0} |E_i| &\le \lim_{\Delta x \to 0} \frac{\{([1 + M\,\Delta x]^{1/(M\,\Delta x)})^{M(b-a)} - 1\}N\,\Delta x}{2M} \\ &= \frac{\{e^{M(b-a)} - 1\}\cdot N \cdot 0}{2M} \\ &= 0, \end{aligned}$$

and the theorem is proved.

6.6. A Runge–Kutta Method

In Sections 6.4 and 6.5 we developed and studied a simple numerical procedure for initial-value problems and provided mathematical support for the method in Theorems 6.2 and 6.3. In this section we will show how to increase the accuracy of the Euler method simply by "improving" on difference equation (6.20). The resulting new method will be called a *Runge–Kutta method.*

In place of Euler's formula,

$$\frac{y_{i+1} - y_i}{\Delta x} = f(x_i, y_i), \tag{6.42}$$

consider the more general formula

$$\frac{y_{i+1} - y_i}{\Delta x} = \beta_0 f(x_i, y_i) + \beta_1 f(x_i + \gamma\,\Delta x, y_i + \delta\,\Delta x), \tag{6.43}$$

where β_0, β_1, γ, and δ are parameters to be determined. Of course, if one sets $\beta_0 = 1, \beta_1 = 0$, then (6.43) reduces to (6.42). The motivation for considering (6.43) lies in the fact that $f(x,y)$ is known for *all* x in $x_i \le x \le x_{i+1}$ and for *all* y, but *only one pair of values* (x,y) has been utilized in (6.42). In (6.43) *we are attempting to incorporate additional known data* with an aim toward improved accuracy.

In an equivalent form, (6.43) can be rewritten as

$$y_{i+1} = y_i + \beta_0(\Delta x) f(x_i, y_i) + \beta_1(\Delta x) f(x_i + \gamma\,\Delta x, y_i + \delta\,\Delta x). \tag{6.44}$$

If f can be written as a Taylor expansion in two variables, then

$$\begin{aligned} &f(x_i + \gamma\,\Delta x, y_i + \delta\,\Delta x) \\ &\quad = f(x_i,y_i) + \gamma(\Delta x) f_x(x_i,y_i) + \delta(\Delta x) f_y(x_i,y_i) \\ &\qquad + \tfrac{1}{2}[\gamma^2(\Delta x)^2 f_{xx}(x_i,y_i) + 2\gamma\delta(\Delta x)^2 f_{xy}(x_i,y_i) + \delta^2(\Delta x)^2 f_{yy}(x_i,y_i)] \\ &\qquad + O((\Delta x)^3), \end{aligned}$$

which, upon substitution into (6.44) and recombination, implies that

$$\begin{aligned} y_{i+1} = y_i &+ \Delta x[(\beta_0 + \beta_1) f(x_i,y_i)] \\ &+ \frac{(\Delta x)^2}{2}[2\beta_1\gamma f_x(x_i,y_i) + 2\beta_1\delta f_y(x_i,y_i)] \\ &+ \frac{(\Delta x)^3}{6}[3\beta_1\gamma^2 f_{xx}(x_i,y_i) + 6\beta_1\gamma\delta f_{xy}(x_i,y_i) + 3\beta_1\delta^2 f_{yy}(x_i,y_i)] \\ &+ \beta_1 O((\Delta x)^4). \end{aligned} \tag{6.45}$$

Suppose also that the exact solution of (6.6) and (6.7) can be written as the Taylor series

$$y_{i+1} = y_i + (\Delta x)y_i' + \frac{(\Delta x)^2}{2}y_i'' + \frac{(\Delta x)^3}{6}y_i''' + \cdots. \tag{6.46}$$

From

$$y' = f(x,y),$$

it follows that

$$y'' = f_x + f_y f,$$

$$y''' = f_{xx} + 2ff_{xy} + f^2 f_{yy} + f_x f_y + ff_y^2,$$

which, upon substitution into (6.46) and recombination, yields

$$\begin{aligned} y_{i+1} = y_i &+ (\Delta x)f(x_i,y_i) + \frac{(\Delta x)^2}{2}[f_x(x_i,y_i) + f_y(x_i,y_i)f(x_i,y_i)] \\ &+ \frac{(\Delta x)^3}{6}\{f_{xx}(x_i,y_i) + 2f(x_i,y_i)f_{xy}(x_i,y_i) + f_{yy}(x_i,y_i)[f(x_i,y_i)]^2 \\ &+ f_x(x_i,y_i)f_y(x_i,y_i) + f(x_i,y_i)[f_y(x_i,y_i)]^2\} + O((\Delta x)^4). \end{aligned} \tag{6.47}$$

We now examine the ways in which we can choose the parameters β_0, β_1, γ, and δ in (6.45) so that expansion (6.45) is in some sense an approximation to (6.47).

For the special choice $\beta_0 = 1$, $\beta_1 = 0$, (6.45) agrees with (6.47) up to, but not including, the $(\Delta x)^2$ terms. This choice, as noted previously, yields the Euler formula (6.20). If we wish to construct a formula for which (6.45) agrees with (6.47) through at least the $(\Delta x)^2$ terms, then setting corresponding coefficients in (6.45) and (6.47) equal implies that

$$\begin{aligned} \beta_0 + \beta_1 &= 1, \\ 2\beta_1\gamma &= 1, \\ 2\beta_1\delta &= f(x_i,y_i). \end{aligned} \tag{6.48}$$

This system of three equations in four unknowns has an infinite number of solutions, a very convenient one of which is

$$\beta_0 = \beta_1 = \tfrac{1}{2}, \qquad \gamma = 1, \quad \delta = f(x_i,y_i). \tag{6.49}$$

Substitution of these parameter values into (6.44) yields

$$y_{i+1} = y_i + \frac{\Delta x}{2}f(x_i,y_i) + \frac{\Delta x}{2}f(x_{i+1},y_i + (\Delta x)f(x_i,y_i)), \tag{6.50}$$

which is called a Runge–Kutta formula and, in the sense developed above, is a higher-order approximation to (6.6) than is the Euler formula (6.20).

Example. Consider the initial-value problem

$$y' = -y + x, \quad y(0) = 1, \qquad 0 \leq x \leq 1.$$

For $\Delta x = 0.1$ and $y_0 = 1$, the numerical solution generated by (6.50) is

$$\begin{aligned} &y_0 = 1, && y_1 = 0.910, && y_2 = 0.838, && y_3 = 0.782,\\ &y_4 = 0.742, && y_5 = 0.714, && y_6 = 0.699, && y_7 = 0.694,\\ &y_8 = 0.700, && y_9 = 0.714, && y_{10} = 0.737, \end{aligned}$$

which is clearly a better approximation to the exact solution (6.28) than is the Euler approximation (6.26).

Note finally that in practice (6.50) is usually written in the more convenient form

$$\begin{aligned} y_{i+1} &= y_i + \tfrac{1}{2}(k_0 + k_1),\\ k_0 &= (\Delta x) f(x_i, y_i),\\ k_1 &= (\Delta x) f(x_{i+1}, y_i + k_0). \end{aligned} \tag{6.51}$$

6.7. Higher-Order Runge–Kutta Formulas

In general, a Runge–Kutta difference approximation of

$$y' = f(x,y) \tag{6.52}$$

on $x_i \leq x \leq x_{i+1}$ is a difference equation of the form

$$y_{i+1} = y_i + \Delta x[\alpha_1 f(\xi_1, \eta_1) + \alpha_2 f(\xi_2, \eta_2) + \cdots + \alpha_k f(\xi_k, \eta_k)], \tag{6.53}$$

where $x_i \leq \xi_j \leq x_{i+1}$, $j = 1, 2, \ldots, k$. Of course, special cases of (6.53) are (6.20) and (6.51). A large number of other Runge–Kutta formulas have been developed and applied. For example, just as (6.51) approximates (6.46) through $(\Delta x)^2$ terms, so it was shown by Heun (1900) that

$$\begin{aligned} y_{i+1} &= y_i + \tfrac{1}{4}(k_0 + 3k_2),\\ k_0 &= (\Delta x) f(x_i, y_i),\\ k_1 &= (\Delta x) f\left(x_i + \frac{\Delta x}{3}, y_i + \frac{k_0}{3}\right),\\ k_2 &= (\Delta x) f\left(x_i + \frac{2\,\Delta x}{3}, y_i + \frac{2k_1}{3}\right) \end{aligned} \tag{6.54}$$

approximates (6.46) through $(\Delta x)^3$ terms, while it was shown by Kutta (1901) that

(6.55)
$$\begin{aligned} y_{i+1} &= y_i + \tfrac{1}{6}(k_0 + 2k_1 + 2k_2 + k_3), \\ k_0 &= (\Delta x)f(x_i, y_i), \\ k_1 &= (\Delta x)f\left(x_i + \frac{\Delta x}{2}, y_i + \frac{k_0}{2}\right), \\ k_2 &= (\Delta x)f\left(x_i + \frac{\Delta x}{2}, y_i + \frac{k_1}{2}\right), \\ k_3 &= (\Delta x)f(x_{i+1}, y_i + k_2) \end{aligned}$$

approximates (6.46) through $(\Delta x)^4$ terms.

6.8. Approximate Solution of Initial-Value Problems for Second-Order Differential Equations

In practice, each of (6.51), (6.54), and (6.55) can be extended to systems of equations. Because of its relative simplicity and high order of approximation, we shall consider only Kutta's formula (6.55). The others can be treated similarly.

For a system of equations of the form

(6.56)
$$\frac{dy}{dx} = F(x,y,v),$$

(6.57)
$$\frac{dv}{dx} = G(x,y,v),$$

we shall take as difference analogues, in direct analogy with (6.55), the Kutta formulas

(6.58)
$$y_{i+1} = y_i + \tfrac{1}{6}(k_0 + 2k_1 + 2k_2 + k_3),$$

(6.59)
$$v_{i+1} = v_i + \tfrac{1}{6}(m_0 + 2m_1 + 2m_2 + m_3),$$

where

$$\begin{aligned} k_0 &= (\Delta x)F(x_i, y_i, v_i), \\ k_1 &= (\Delta x)F\left(x_i + \frac{\Delta x}{2}, y_i + \frac{k_0}{2}, v_i + \frac{m_0}{2}\right), \\ k_2 &= (\Delta x)F\left(x_i + \frac{\Delta x}{2}, y_i + \frac{k_1}{2}, v_i + \frac{m_1}{2}\right), \end{aligned}$$

$$k_3 = (\Delta x)F(x_{i+1}, y_i + k_2, v_i + m_2),$$

$$m_0 = (\Delta x)G(x_i, y_i, v_i),$$

$$m_1 = (\Delta x)G\left(x_i + \frac{\Delta x}{2}, y_i + \frac{k_0}{2}, v_i + \frac{m_0}{2}\right),$$

$$m_2 = (\Delta x)G\left(x_i + \frac{\Delta x}{2}, y_i + \frac{k_1}{2}, v_i + \frac{m_1}{2}\right),$$

$$m_3 = (\Delta x)G(x_{i+1}, y_i + k_2, v_i + m_2).$$

If we return now, at long last, to (6.1), that is, to

$$y'' = f(x,y,y') \tag{6.60}$$

and make the change of variables

$$y' = v$$

so that (6.60) is equivalent to the system

$$\frac{dy}{dx} = v, \tag{6.61}$$

$$\frac{dv}{dx} = f(x,y,v), \tag{6.62}$$

then the Kutta formulas reduce to

$$y_{i+1} = y_i + (\Delta x)v_i + \frac{\Delta x}{6}(m_0 + m_1 + m_2), \tag{6.63}$$

$$v_{i+1} = v_i + \tfrac{1}{6}(m_0 + 2m_1 + 2m_2 + m_3), \tag{6.64}$$

where

$$m_0 = (\Delta x)f(x_i, y_i, v_i), \tag{6.65}$$

$$m_1 = (\Delta x)f\left(x_i + \frac{\Delta x}{2}, y_i + \frac{v_i\,\Delta x}{2}, v_i + \frac{m_0}{2}\right), \tag{6.66}$$

$$m_2 = (\Delta x)f\left(x_i + \frac{\Delta x}{2}, y_i + \frac{v_i\,\Delta x}{2} + \frac{m_0\,\Delta x}{4}, v_i + \frac{m_1}{2}\right), \tag{6.67}$$

$$m_3 = (\Delta x)f\left(x_{i+1}, y_i + v_i\,\Delta x + \frac{m_1\,\Delta x}{2}, v_i + m_2\right). \tag{6.68}$$

For an initial-value problem for (6.60), one knows y_0 and v_0. These are used first in (6.65) through (6.68) to determine m_0, m_1, m_2, m_3, which in turn are used to generate y_1 from (6.63) and v_1 from (6.64). Substitution then of y_1 and v_1 into (6.65) through (6.68) generates new values m_0, m_1, m_2, m_3, which are then used to generate y_2 from (6.63) and v_2 from (6.64). And so the process continues and one generates $y_3, y_4, \ldots, y_n$, in the indicated recursive fashion.

Rather than illustrate the method with a trivial problem, however, we shall this time consider a problem of long-standing interest to engineers, mathematicians, and physicists and show how to solve it numerically by means of (6.63) through (6.68).

6.9. Nonlinear Pendulum

Consider a pendulum system that is hinged at O and has mass m centered at P, as shown in Figure 6.1. We assume that the length $|OP|$ is l, that θ is the angular measure in radians of the pendulum's deviation from the vertical OQ, and that P is constrained to move on the circle whose center is O and whose radius is l. The problem to be considered is that of describing the motion of P after its release from a given initial position at which it was at rest.

Physically, it is known that the motion of the pendulum damps out and that the length of time between each swing decreases with time.

Analytically, we proceed as follows. The motion of P is assumed to be governed by Newton's dynamical equation

$$F = m \cdot a. \tag{6.69}$$

Since circular arc QP has length $l\theta$, it follows that

$$a = \frac{d^2}{dt^2}(l\theta) = l\ddot{\theta},$$

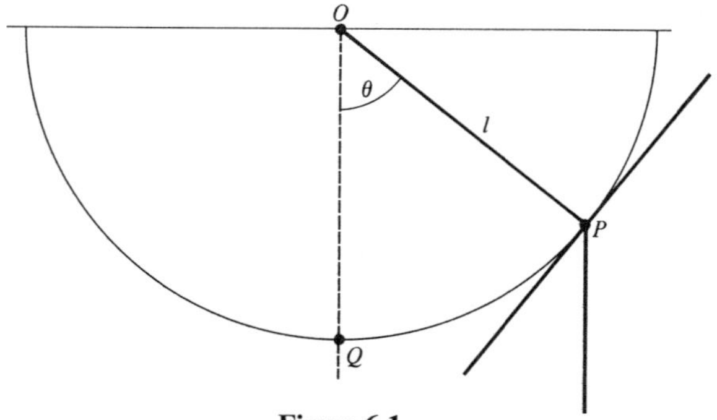

Figure 6.1

so (6.69) becomes

(6.70) $$ml\ddot{\theta} = F.$$

In considering the forces that act on P, let F_1 be a gravitational component and let F_2 be a damping component. Then

(6.71) $$F_1 = -mg, \qquad g > 0$$

and let us take F_2 in the form

(6.72) $$F_2 = -c\dot{\theta}, \qquad c > 0.$$

But, since the component of F_1 that is tangent to the circle at P is $-mg \sin \theta$, it follows that

$$F = -mg \sin \theta - c\dot{\theta}.$$

Thus (6.70) becomes

$$ml\ddot{\theta} = -mg \sin \theta - c\dot{\theta}$$

or, equivalently,

(6.73) $$\ddot{\theta} + \frac{c}{ml}\dot{\theta} + \frac{g}{l} \sin \theta = 0.$$

Our problem, then, is that of solving (6.73) subject to the given initial conditions

(6.74) $$\theta(0) = \alpha, \qquad \dot{\theta}(0) = 0.$$

Example. Describe the pendulum motion defined by

(6.75) $$\ddot{\theta} + (0.3)\dot{\theta} + \sin \theta = 0,$$

(6.76) $$\theta(0) = \frac{\pi}{4}, \qquad \dot{\theta}(0) = 0.$$

Solution. Since there is no known analytical method for solving nonlinear initial-value problem (6.75) and (6.76), the problem was solved numerically by setting $y = \theta$, $x = t$, and by applying (6.63) through (6.68) with $\Delta t = 0.01$. Computation was executed on the UNIVAC 1108 in double precision for 15,000 time steps, that is, for 150 seconds of pendulum oscillation, with a total machine computing time of 2 seconds. The first 25 seconds of pendulum oscillations are shown graphically in Figure 6.2, where the peak values 0.75462, -0.47647, 0.29335, -0.18156, 0.11259, -0.06987, 0.04337, and -0.02693 occurred at the respective times 0, 3.28, 6.49, 9.68, 12.86, 16.04, 19.22, and 22.40. The time required for the particle to travel from one peak

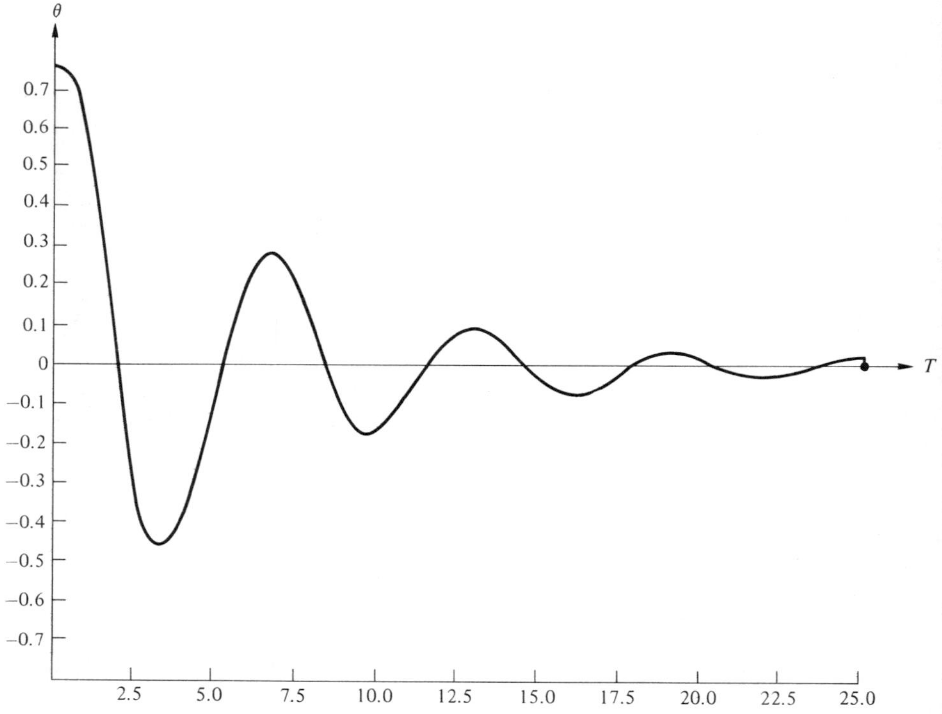

Figure 6.2

to another decreased monotonically and damping was present during the entire 150 seconds of motion.

The pendulum motion described in the above example agrees with what happens in a real physical environment. Any attempt to substitute the linear approximation

$$\sin\theta = \theta$$

into (6.75) so that one can use analytical methods to solve the problem leads to results that do not agree with experiment. If, for example, one first also eliminates the damping factor, so that (6.75) and (6.76) reduce to

$$\ddot{\theta} + \theta = 0, \tag{6.77}$$

$$\theta(0) = \frac{\pi}{4}, \qquad \dot{\theta}(0) = 0, \tag{6.78}$$

then the solution

$$\theta = \frac{\pi}{4}\cos t$$

of (6.77) and (6.78) exhibits neither damping nor a monotonic decrease in the time interval between consecutive peak values. Retention of the damping factor so that (6.75) and (6.76) reduces to

$$\ddot{\theta} + (0.3)\dot{\theta} + \theta = 0, \tag{6.79}$$

$$\theta(0) = \frac{\pi}{4}, \qquad \dot{\theta}(0) = 0, \tag{6.80}$$

leads, by (4.17), to the solution

$$\theta = \frac{(0.15)\pi}{4(0.9775)^{1/2}}e^{(-0.15)t}\sin[(0.9775)^{1/2}t] + \frac{\pi}{4}e^{(-0.15)t}\cos[(0.9775)^{1/2}t] \tag{6.81}$$

or, equivalently, to

$$\theta = \left[\left(\frac{\pi}{4}\right)^2 + \left(\frac{0.15\pi}{4(0.9775)^{1/2}}\right)^2\right]^{1/2} e^{(-0.15)t}\sin[\phi + (0.9775)^{1/2}t], \tag{6.82}$$

where ϕ is that acute angle which satisfies

$$\sin\phi = \frac{\pi/4}{\{(\pi/4)^2 + [0.15\pi/(4(0.9775)^{1/2})]^2\}^{1/2}}.$$

Although (6.82) does possess the desired damping, it does not exhibit a decrease in the interval time between consecutive peaks due to the periodicity of the sine component of the solution. The constant time interval between successive passes through Q is $\pi/\sqrt{0.9775}$, or approximately 3.18 seconds. This agrees to two decimal places with the results of the numerical example above for all $t \geq 9.68$.

6.10. Instability

From the computing point of view, a set of calculations that results in overflow is called *unstable*. Otherwise, a computation is called *stable*. When instability occurs, one must first check for program and machine errors. If neither of these is present, then one must apply analytical methods to the mathematical algorithm to find the source of trouble. Two of the more common mathematical sources of instability will now be discussed by means of illustrative examples.

Consider first the initial-value problem

(6.83) $$y' = -100y, \qquad y(0) = 1,$$

the exact solution of which is

(6.84) $$y = e^{-100x}.$$

The solution (6.84) is always positive and tends to zero as x goes to infinity. Suppose, now, one applies Euler's method with $\Delta x = 0.1$ to approximate a solution of (6.83). Then, from

$$y_{i+1} = -9y_i, \qquad i = 0, 1, \ldots,$$
$$y_0 = 1,$$

one generates the numerical solution

$$y_1 = -9, y_2 = (-9)^2, y_3 = (-9)^3, y_4 = (-9)^4, \ldots, y_n = (-9)^n, \ldots,$$

which, in short order, results in computer overflow. In order to analyze the instability, we proceed as follows. For arbitrary Δx, application of Euler's method to (6.83) yields

$$y_{i+1} = y_i - 100(\Delta x)y_i = y_i(1 - 100\Delta x).$$

Thus

$$\begin{aligned} y_1 &= y_0(1 - 100\Delta x) = (1 - 100\Delta x), \\ y_2 &= y_1(1 - 100\Delta x) = (1 - 100\Delta x)^2, \\ y_3 &= (1 - 100\Delta x)^3, \\ &\vdots \\ y_n &= (1 - 100\Delta x)^n. \end{aligned}$$

For instability, the sequence $|y_n|$, $n = 0, 1, 2, \ldots,$ must exceed the largest number in one's computer, which, assuming that all computations are exact, can only happen if $|1 - 100\Delta x| > 1$. Thus, to assure stability, one must choose Δx to satisfy $|1 - 100\Delta x| \le 1$, or, equivalently, $\Delta x \le \frac{1}{50}$. Thus, for example, for $\Delta x = 0.001$, Euler's formula becomes

$$y_{i+1} = (0.9)y_i, \qquad i = 0, 1, \ldots,$$

and iteration with $y_0 = 1$ yields

$$y_1 = (0.9), y_2 = (0.9)^2, y_3 = (0.9)^3, y_4 = (0.9)^4, \ldots, y_n = (0.9)^n.$$

Calculation with $\Delta x = \frac{1}{50}$, although stable, yields a far less desirable numerical solution than the one with $\Delta x = 0.001$.

Consider next a source of instability that is somewhat different than the one above. Suppose one is asked to solve the second-order difference equation

(6.85) $$y_{i+2} - \tfrac{5}{2}y_{i+1} + y_i = 0, \qquad i = 0, 1, \ldots, n$$

with given initial conditions

(6.86) $$y_0 = \tfrac{1}{2}, \qquad y_1 = \tfrac{1}{4}.$$

Then, in a straightforward fashion, one has, from (6.85) and (6.86),

(6.87) $$y_{i+2} = \tfrac{5}{2}y_{i+1} - y_i, \qquad i = 0, 1, \ldots,$$

and

$$\begin{aligned}
y_2 &= \tfrac{5}{2}y_1 - y_0 = \tfrac{5}{2}(\tfrac{1}{4}) - \tfrac{1}{2} = \tfrac{1}{8},\\
y_3 &= \tfrac{5}{2}y_2 - y_1 = \tfrac{5}{2}(\tfrac{1}{8}) - \tfrac{1}{4} = \tfrac{1}{16},\\
y_4 &= \tfrac{5}{2}y_3 - y_2 = \tfrac{5}{2}(\tfrac{1}{16}) - \tfrac{1}{8} = \tfrac{1}{32},\\
y_5 &= \tfrac{5}{2}y_4 - y_3 = \tfrac{5}{2}(\tfrac{1}{32}) - \tfrac{1}{16} = \tfrac{1}{64},\\
&\vdots\\
y_n &= (\tfrac{1}{2})^{n+1},
\end{aligned}$$

and it follows easily that $y_n \to 0$ as $n \to \infty$. Now suppose that instead of doing exact calculations, as above, one simulates what a digital computer does and allows roundoff error to be introduced in the following, ultrasimple way. Suppose that in calculating y_i one simply *rounds off input and computations to one decimal place.* Thus initially one has

(6.88) $$y_0 = 0.5,$$

(6.89) $$y_1 = 0.3$$

and, from (6.85),

$$\begin{aligned}
y_2 &= \tfrac{5}{2}y_1 - y_0 = \tfrac{5}{2}(0.3) - (0.5) = (2.5)(0.3) - (0.5)\\
&= 0.8 - 0.5 = 0.3,\\
y_3 &= \tfrac{5}{2}y_2 - y_1 = \tfrac{5}{2}(0.3) - (0.3) = 0.5,\\
y_4 &= \tfrac{5}{2}y_3 - y_2 = \tfrac{5}{2}(0.5) - (0.3) = 1.0,\\
y_5 &= \tfrac{5}{2}y_4 - y_3 = \tfrac{5}{2}(1.0) - (0.5) = 2.0,\\
y_6 &= \tfrac{5}{2}y_5 - y_4 = \tfrac{5}{2}(2.0) - (1.0) = 4.0,\\
y_7 &= \tfrac{5}{2}y_6 - y_5 = \tfrac{5}{2}(4.0) - (2.0) = 8.0,\\
&\vdots\\
y_n &= 2^{n-4}, \qquad n \geq 4,
\end{aligned}$$

which, again, in short order yields overflow on any computer. The analysis of the instability this time, however, requires greater care. For this purpose, let

$$(6.90) \qquad y_i = \lambda^i, \qquad \lambda \text{ a constant}, \quad i = 1, 2, \ldots,$$

be substituted into (6.85), so that

$$\lambda^{i+2} - \tfrac{5}{2}\lambda^{i+1} + \lambda^i = 0.$$

If $\lambda \neq 0$, this last equation implies that

$$2\lambda^2 - 5\lambda + 2 = 0,$$

whose two roots are

$$\lambda_1 = 2, \qquad \lambda_2 = \tfrac{1}{2}.$$

In analogy with second-order linear differential equations, one has that

$$(6.91) \qquad y_i = c_1 2^i + c_2(\tfrac{1}{2})^i$$

is a solution of (6.85) for any constants c_1 and c_2.

Now, conditions (6.86) imply that

$$c_1 = 0, \qquad c_2 = \tfrac{1}{2}.$$

so that (6.91) reduces to

$$(6.92) \qquad y_i = (\tfrac{1}{2})^{i+1}, \qquad i = 0, 1, 2, \ldots.$$

Formula (6.92) yields the well-behaved solution first generated by using exact calculation. But when one used the rounding procedure described above so that the initial conditions are (6.88) and (6.89), then one has, from these conditions, that

$$(6.93) \qquad y_i = \tfrac{1}{30}(2^i) + \tfrac{7}{15}(\tfrac{1}{2})^i.$$

The presence of the term $\frac{1}{30}(2^i)$ in (6.93) causes our overflow as i gets large and is a result of using a rounding procedure, which is, of course, inherent in one form or another in every computer.

Thus far we have exhibited two unstable calculations, one that can actually be corrected by choosing Δx sufficiently small, the other which can never be eliminated because of the process of rounding numbers in actual machine computation. There are other sources of instability and the subject is one of extensive study. The interested reader is encouraged to examine such topics as pointwise, stepwise, local, strong, and weak stability, and the intimate relationship between stability and convergence in such texts as Hildebrand (3) and Isaacson and Keller.

6.11. Remarks

There exist a large number of other methods which can be of value in the numerical solution of initial-value problems. General categories include numerical integration, predictor–corrector, continuous, and multistep methods. There also exist valuable special methods for special problems, such as those relating to stiff equations, van der Pol's equation, and second-order equations that do not contain y'. Finally, note that often one can conveniently solve a boundary-value problem by an initial-value technique and an initial-value problem by a boundary-value technique.

General reading sources for the material outlined above are Collatz (1), Conte, Fehlberg, Fox (1, 2), Froberg, Goodwin, Greenspan (4, 6), Hamming, Henrici (1), Hildebrand (2, 3), Kelly, Milne (2), Noble, Ralston, Ralston and Wilf, Sarafyan, and Todd.

EXERCISES

1. Reduce the following second-order initial-value problems to systems:
   (a) $y'' - 3y' - 4y = 7x + 7$, $y(0) = 1$, $y'(0) = 2$
   (b) $y'' - 3xyy' = 0$, $y(0) = 1$, $y'(0) = -3$
   (c) $y'' = e^{xy'}$, $y(0) = 4$, $y'(0) = 2$
   (d) $y'' = e^{xy}$, $y(0) = -6$, $y'(0) = 3$
   (e) $y'' - 7(1 - y^2)y' - 6y = 0$, $y(0) = 1$, $y'(0) = 0$
2. Find the exact solution of each of the following initial-value problems:
   (a) $y' = x$, $y(0) = 1$ (b) $y' = y - x$, $y(0) = -1$
   (c) $y' = xy - 2x$, $y(0) = 4$ (d) $y' = x^2y$, $y(0) = 1$
   (e) $y' = (\sin x)y - x$, $y(0) = 2$
3. Let R_5 be the set $x_0 = 0, x_1 = 1, x_2 = 2, x_3 = 3, x_4 = 4$. On R_5 find the solution of each of the following initial-value problems:
   (a) $y_0 = 1$; $y_{i+1} = (y_i)^2 + (x_i)^2$, $i = 0, 1, 2, 3$
   (b) $y_0 = -1$; $y_{i+1} = (y_i)^2 + (\Delta x)(x_i)^2$, $i = 0, 1, 2, 3$
   (c) $y_0 = 0$; $y_{i+1} = (y_i)^2(x_i)^2$, $i = 0, 1, 2, 3$
   (d) $y_0 = 3$; $y_{i+1} = -1/y_i$, $i = 0, 1, 2, 3$
4. Solve each of the following initial-value problems on $0 \le x \le 1$ by Euler's method with $\Delta x = 0.2$:
   (a) $y' = x$, $y(0) = 1$ (b) $y' = y - x$, $y(0) = -1$
   (c) $y' = xy - 2x$, $y(0) = 4$ (d) $y' = y^2$, $y(0) = 1$
   (e) $y' = xy^2 - 2y - x^3$, $y(0) = 1$ (f) $y' = e^{-xy}$, $y(0) = 0$
5. State the conclusions that can be drawn from Lemma 6.1 in each of the following cases:
   (a) $|E_{i+1}| \le |E_i|$, $i = 0, 1, \ldots, n - 1$
   (b) $|E_{i+1}| \le |E_i| + 1$, $i = 0, 1, \ldots, n - 1$

(c) $|E_{i+1}| \leq |E_i| - 2, i = 0, 1, \ldots, n-1$
(d) $|E_{i+1}| \leq 2|E_i| + 3, i = 0, 1, \ldots, n-1$

6. For each of the following, estimate the constant M of Theorem 6.2:
   (a) $y' = x + y, 0 < x < 1$ (b) $y' = xy, 0 < x < 2$
   (c) $y' = x + \sin y, 0 < x < 2$ (d) $y' = x^2 \cos y, 1 < x < 3$
   (e) $y' = x^2 - \cos y, 0 < x$
7. Why does the first inequality in the proof of Theorem 6.3 follow from (6.41)?
8. Solve on $0 \leq x \leq 1$ by the Runge–Kutta method of Section 6.6 with $\Delta x = 0.2$. Compare your results with those of Exercise 4.
   (a) $y' = x, y(0) = 1$ (b) $y' = y - x, y(0) = -1$
   (c) $y' = xy - 2x, y(0) = 4$ (d) $y' = y^2, y(0) = 1$
   (e) $y' = xy^2 - 2y - x^3, y(0) = 1$
9. Derive Heun's formula (6.54).
10. Derive Kutta's formula (6.55).
11. Derive Kutta's formulas for $y'' = f(x,y,y')$ from his formulas for the system $y' = F(x,y,v), v' = G(x,y,v)$.
12. Apply Heun's and Kutta's formulas to $y' = -y + x, y(0) = 1, 0 \leq x \leq 1$, $\Delta x = 0.1$, and compare the results with the exact solution and with the numerical results of Sections 6.4 and 6.6.
13. Solve each of the following only at $x_0, x_1, x_2, x_3, x_4, x_5$ by Kutta's formulas with $\Delta x = 0.01$:
   (a) $y'' + (0.3)y' + \sin y = 0, y(0) = \pi/4, y'(0) = 0$
   (b) $y'' + \sin y = 0, y(0) = \pi/2, y'(0) = 0$
   (c) $y'' + (0.04)y' + \sin y = 0, y(0) = \pi, y'(0) = 0$
14. Describe the pendulum motion determined by each of the following:
   (a) $\ddot{\theta} = 0, \theta(0) = 0, \dot{\theta}(0) = 0$
   (b) $\ddot{\theta} = 0, \theta(0) = \pi/2, \dot{\theta}(0) = 0$
   (c) $\ddot{\theta} = 0, \theta(0) = \pi, \dot{\theta}(0) = 0$
   (d) $\ddot{\theta} + \theta = 0, \theta(0) = \pi, \dot{\theta}(0) = 0$
   (e) $\ddot{\theta} + (0.3)\dot{\theta} + \theta = 0, \theta(0) = \pi/2, \dot{\theta}(0) = 0$
   (f) $\ddot{\theta} + (0.3)\dot{\theta} + \sin\theta = 0, \theta(0) = \pi/2, \dot{\theta}(0) = 0$
   (g) $\ddot{\theta} + (0.3)\dot{\theta} + \sin\theta = 0, \theta(0) = \pi, \dot{\theta}(0) = 0$
   (h) $\ddot{\theta} + (0.3)\dot{\theta} + \sin\theta = 0, \theta(0) = 99\pi/100, \dot{\theta}(0) = 0$
   (i) $\ddot{\theta} + (0.1)\dot{\theta} + \sin\theta = 0, \theta(0) = \pi/4, \dot{\theta}(0) = 0$
   (j) $\ddot{\theta} + (0.0001)\dot{\theta} + \sin\theta = 0, \theta(0) = \pi/4, \dot{\theta}(0) = 0$
15. Euler's method is applied to $y' = -10y, y(0) = 1$, with exact calculations for each of (a) $\Delta x = 0.1$, (b) $\Delta x = 0.01$, (c) $\Delta x = 0.001$. Determine which of the computations are stable and which are unstable.
16. Each of the following difference equations is solved numerically with *all* calculations rounded to two decimal places.

(a) $y_{i+2} - (\frac{5}{2})y_{i+1} + y_i = 0$, $y_0 = \frac{1}{2}$, $y_1 = \frac{1}{4}$
(b) $y_{i+2} + 3y_{i+1} + 2y_i = 0$, $y_0 = 0$, $y_1 = 0$
(c) $y_{i+2} + 3y_{i+1} + 2y_i = 0$, $y_0 = 0.01$, $y_1 = 0.00$
(d) $4y_{i+2} - y_i = 0$, $y_0 = 0$, $y_1 = 0$
(e) $4y_{i+2} - y_i = 0$, $y_0 = 0.01$, $y_1 = 0.00$

Determine which of the computations are stable and which are unstable.

17. Consider the Runge–Kutta formula (6.50) in the form

$$y_{i+1}^{(1)} = y_i + (\Delta x) f(x_i, y_i),$$

$$y_{i+1} = y_i + \frac{\Delta x}{2}[f(x_i, y_i) + f(x_{i+1}, y_{i+1}^{(1)})].$$

The first equation can be interpreted as an initial approximation, or prediction, by Euler's formula, for y_{i+1}. If the second equation is rewritten as an iteration formula in the form

$$y_{i+1}^{(k+1)} = y_i + \frac{\Delta x}{2}[f(x_i, y_i) + f(x_{i+1}, y_{i+1}^{(k)})], \qquad k = 1, 2, 3, \ldots,$$

then it is called a *corrector formula*. Iteration with the corrector formula continues until $y_{i+1}^{(m)} = y_{i+1}^{(m+1)}$ for some value m, and one sets $y_{i+1} = y_{i+1}^{(m)}$. The resulting method, performed for each value of i, is called a *predictor–corrector method*. Apply, then, the above predictor–corrector method to the initial-value problems of Exercise 4.

Chapter 7

DISCRETE MODEL THEORY

7.1. Introduction

Discrete model theory is concerned with the description of physical concepts and laws by means of finite sets rather than by infinite ones. Discrete functions and difference equations play fundamental roles in such an approach.

In this chapter we shall illustrate ideas and methods of discrete model theory by developing basic discrete, plane Newtonian dynamics, thereby constructing a physical theory that will be compatible with digital computer capabilities. Such an approach to dynamics has been of occasional interest to mathematicians throughout the centuries, but only with the development of modern computational machinery has it become a source of exceptional power. The equations of our model, whether linear or nonlinear, will be easily solvable by means only of arithmetic processes, and the physically nonrealizable concept of infinity and convergence proofs required for continuous models will become unnecessary. Finally, it is worth noting that the existence and uniqueness of the solution of a dynamics problem will usually be a trivial consequence of the recursive structure of the associated difference equation. Thereby we shall have constructed a simplified mathematical approach that avoids the topological theory necessary for the existence and uniqueness theorems of continuous mathematics.

It will be important in our development that we carefully *avoid* being motivated by well-known, commonly accepted, continuous physical concepts if we are, in fact, to construct a truly simpler mathematical approach to dynamics, and this will be done.

7.2. Implications of Constraints on Measurement

In the simplest of all possible worlds, experimental scientists use three basic measuring instruments, one that measures time and will be called a *clock*, one that measures weight and will be called a *scale*, and one that measures length and will be called a *ruler*. Each instrument has limited accuracy and in practice is applied only a finite number of times to produce a finite set of data. Often these data are combined by arithmetic processes to determine certain additional scalar quantities, such as speed or work. At other times one may use the data to infer certain equations associated with the system under consideration and then proceed to solve the equations in some comprehensive way. In any event, a clock, scale, and ruler will always be considered as part of a given experiment and hence define a smallest meaningful time interval (which is positive), a smallest meaningful weight (which is positive), and a smallest meaningful length (which is positive). These smallest positive units are determined by the precision of the clock, scale, and ruler, and no smaller units are meaningful because they cannot be measured by the given instruments.

By necessity, then, the data of a given experiment consist of rational numbers that have a finite number of significant digits, and it is important to note that the storage of and arithmetic combination of such numbers is the primary function of the high-speed digital computer. It is for this reason that discrete models will be more intimately associated with experimental, rather than theoretical, science.

7.3. Special Discrete Functions

Of special interest from the digital computer point of view are the rational, square-root, and trigonometric discrete functions, which will be defined as follows.

A *rational* discrete function is one whose values are determined from an independent variable by a finite number of arithmetic operations (where, of course, division by zero is *not* an arithmetic operation).

Example 1. On the R_{11} set $x_i = i/10, i = 0, 1, \ldots, 10$, each of the following is a discrete rational function:

$$5, x, 1 + x, 1 - x, \frac{2}{1 + x}, \frac{x}{1 + x^2}, \frac{1 - x}{1 + x^2}, \frac{1 - 2x + 3x^2}{1 + x + x^2 + x^6}.$$

If $f(x)$ is a discrete, *nonnegative* rational function, then the discrete *square root* of f, denoted by $F = f^{1/2}$, is defined as follows. Let the elements x_i of a given R_{n+1} set each contain a given, fixed number of significant digits. Let f be defined on R_{n+1}. Then $[f(x)]^{1/2}$ is that rational number y which has the same number of significant digits as $f(x)$ and whose square best approximates $f(x)$. Of course the actual determination of a square root can be accomplished on a computer by means of the elementary square-root procedure found in almost all books on arithmetic.

Example 2. Let R_6 be the set $x_0 = 0.0$, $x_1 = 0.1$, $x_2 = 0.2$, $x_3 = 0.3$, $x_4 = 0.4$, $x_5 = 0.5$. On R_6 let $f(x) = 1 + x^2$, so that

$$f(0.0) = 1.00,$$
$$f(0.1) = 1.01,$$
$$f(0.2) = 1.04,$$
$$f(0.3) = 1.09,$$
$$f(0.4) = 1.16,$$
$$f(0.5) = 1.25.$$

Then

$$[f(0.0)]^{1/2} = 1.00,$$
$$[f(0.1)]^{1/2} = 1.00,$$
$$[f(0.2)]^{1/2} = 1.02,$$
$$[f(0.3)]^{1/2} = 1.04,$$
$$[f(0.4)]^{1/2} = 1.08,$$
$$[f(0.5)]^{1/2} = 1.12.$$

If, for some computational reason, each $f(x_i)$ above had been given to only two significant digits, then the square roots would also have had to be given to only two significant digits.

With regard to the geometry of a right triangle, it will be assumed that, given measured lengths of any two sides, the law of Pythagoras and the definition above of the square-root function can be applied to yield the length of the third side.

With regard to trigonometric functions, it will be assumed for the present that the accuracy with which one can measure angles is that defined by the ruler of a given experiment. The trigonometric functions themselves will be treated in the usual elementary fashion as ratios of various segments of right triangles, because these definitions involve only rational functions and square roots.

Since the square-root function is defined above only in terms of arithmetic operations, and since trigonometric functions are all ratios, we shall assume that tables of these functions are available for our use.

7.4. Particles, Time, and Motion

From a purely mathematical point of view, one can consider the terms "particle," "time," and "motion" as undefined and then proceed to define other concepts in terms of these. Nevertheless, physically, it is desirable to have some intuition about these rudiments, and our present purpose is to develop such intuition.

For a given experiment, a *particle* of a given solid will be considered to be the smallest spherical portion of the solid whose diameter can be measured to the accuracy of the associated ruler and whose weight can be measured to the accuracy of the associated scale. A plane particle, which is the only kind with which we shall deal here, will be any great circle section of a particle. The centroid of a plane particle is defined to be the center of the associated great circle. By the position of a particle we shall mean the position of its centroid. The mass of a particle is defined in the usual way in Newtonian mechanics. All plane figures are to be considered as compositions of particles.

The concept of *time* that we shall use can be described as follows. Let Δt be the smallest measureable time interval defined by a given clock. On a number axis, mark off the time positions $t_k = k\,\Delta t$, $k = 0, 1, 2, \ldots, n$, where n is a fixed positive integer. The numbers t_0 and t_n may be considered to be initial and terminal times, respectively, of some physical event. The passage of time between t_0 and t_n may be thought of as follows. At first a small "packet" or "particle" of time containing Δt units of time is located at t_0. In Figure 7.1 this packet is denoted by circle C with diameter Δt and center at t_0. After Δt seconds, C is located with its center at t_1, after $2\,\Delta t$ seconds with its center at t_2, after $3\,\Delta t$ seconds with its center at t_3, and so on, until after $n\,\Delta t$ seconds the packet is located with its center at t_n. The passage of time between t_0 and t_n is conceived then only as the presence of time packet C with its center at t_0, $t_1, \ldots, t_n$ at the respective times $0, \Delta t, \ldots, n\,\Delta t$. How packet C got from t_0 to t_n can be idealized as follows. C remains at t_0 (that is, its center remains at t_0) until Δt seconds have elapsed. It then makes an instantaneous jump, called a *quantum jump*, to t_1. It remains at t_1 until Δt additional seconds have elapsed, at which time it makes a second quantum jump and locates at t_2. After n such quantum jumps, the center of C appears at t_n.

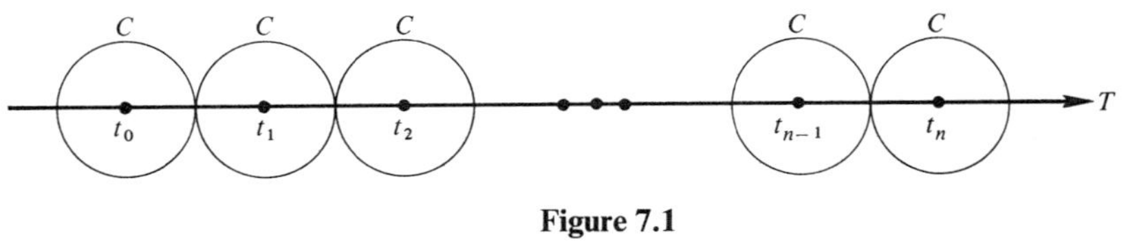

Figure 7.1

That the above model of the passage of time is reasonable is a consequence of the limitation that no time interval of length smaller than Δt can be measured by the given clock.

Let us next examine how the above concept of time applies to the development of a simple concept of one-dimensional *motion*, that is, of motion in a fixed direction. Let Δx be the smallest measurable length defined by a given ruler and let Δt be the smallest measurable time interval defined by a given clock. On an X axis, mark off $x_k = k\,\Delta x, k = 0, 1, \ldots, m$, and on a T axis mark off $t_j = j\,\Delta t, j = 0, 1, \ldots, n$, as shown in Figure 7.2. For illustrative purposes, assume that a particle C has center P which is located at x_0 when $t = t_0$, at x_1 when $t = t_3$, at x_2 when $t = t_6$, and at x_3 when $t = t_{10}$. The motion of C from x_0 to x_3 is merely C's being with center P at x_0, x_1, x_2, x_3 at respective times t_0, t_3, t_6, t_{10}. Thus the motion of C from x_0 to x_3 is considered to be a sequence of four "stills." This concept of motion is biologically acceptable and is realized physically in motion pictures, where the viewer observes motion from a finite sequence of stills projected with sufficient rapidity and transmitted as retinal images to the brain.

Note, incidentally, that the axes and curves drawn in Figures 7.1 and 7.2 consist, *physically*, of a large, but *finite*, number of molecules of paper.

7.5. Velocity and Acceleration

We consider now the basic concepts of velocity and acceleration and, for simplicity, continue to confine our attention to a particle moving in a fixed direction.

For $\Delta t > 0$, let $t_k = k\,\Delta t$, $k = 0, 1, \ldots, q-1, q$. At time t_k, let a particle that is in motion in a fixed X direction, represented in Figure 7.3 by an X axis, have its center at x_k. We wish to define the *velocity* v_k and *acceleration* a_k of the particle at each time t_k, $k = 1, 2, \ldots, q$. For this purpose, consider first the interval $t_0 \le t \le t_1$. Suppose, as shown in Figure 7.3, one knows v_0

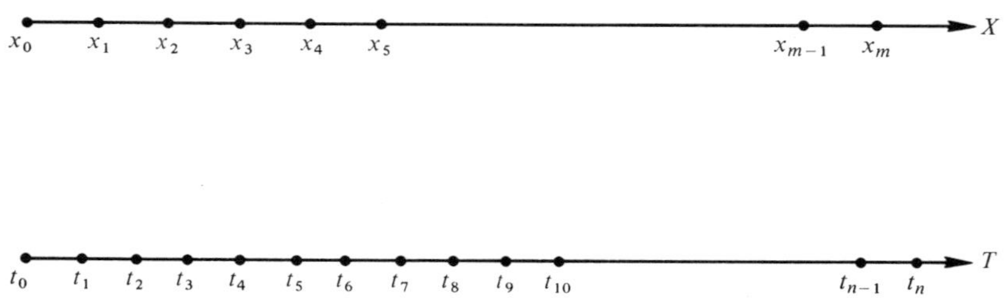

Figure 7.2

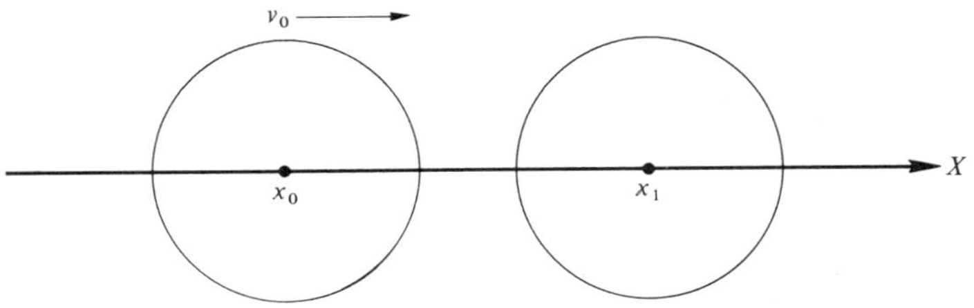

Figure 7.3

in addition to x_0 and x_1. For example, when, in a falling-body problem, a particle's motion begins from a position of rest, one would know that $v_0 = 0$. Let us try to define $v_1 = v(t_1)$ in a fashion that will use *all* the given data. This can be accomplished, for example, by defining v_1 implicitly by the smoothing formula

$$\frac{x_1 - x_0}{\Delta t} = \frac{v_0 + v_1}{2}. \tag{7.1}$$

If one were to define v_1 by the backward-difference formula

$$v_1 = \frac{x_1 - x_0}{\Delta t}, \tag{7.2}$$

then the knowledge of v_0 would have been neglected. If one were to define velocity by a forward-difference formula, it would follow that

$$v_0 = \frac{x_1 - x_0}{\Delta t} \tag{7.3}$$

and $v_0 = 0$ would imply that $x_0 = x_1$. However, it is undesirable physically for such rudimentary problems like those for falling bodies to imply that a particle dropped from a position of rest cannot change its position when t changes from t_0 to t_1.

Thus the above disadvantages of (7.2) and (7.3) lead to the choice of (7.1), which then motivates the general definition

$$\frac{x_k - x_{k-1}}{\Delta t} = \frac{v_{k-1} + v_k}{2}, \qquad k = 1, 2, \ldots, q, \tag{7.4}$$

for velocity v_k. If one writes (7.4) in the equivalent form

$$v_k = 2\frac{x_k - x_{k-1}}{\Delta t} - v_{k-1}, \qquad k = 1, 2, \ldots, q, \tag{7.5}$$

then

$$v_1 = \frac{2}{\Delta t}[x_1 - x_0] - v_0 \tag{7.6}$$

$$v_2 = \frac{2}{\Delta t}[x_2 - x_1] - v_1 = \frac{2}{\Delta t}[x_2 - 2x_1 + x_0] + v_0, \tag{7.7}$$

$$v_3 = \frac{2}{\Delta t}[x_3 - x_2] - v_2 = \frac{2}{\Delta t}[x_3 - 2x_2 + 2x_1 - x_0] - v_0, \tag{7.8}$$

$$v_4 = \frac{2}{\Delta t}[x_4 - x_3] - v_3 = \frac{2}{\Delta t}[x_4 - 2x_3 + 2x_2 - 2x_1 + x_0] + v_0, \tag{7.9}$$

$$\vdots$$

$$v_k = \frac{2}{\Delta t}\left[x_k + (-1)^k x_0 + 2\sum_{j=1}^{k-1}(-1)^j x_{k-j}\right] + (-1)^k v_0, \qquad k \geq 2. \tag{7.10}$$

With regard to acceleration $a_k = a(t_k), k = 1, 2, \ldots, q$, one rarely knows a_0, so it is reasonable to define a_1 by the backward difference

$$a_1 = \frac{v_1 - v_0}{\Delta t}, \tag{7.11}$$

from which we are motivated to define a_k, in general, by

$$a_k = \frac{v_k - v_{k-1}}{\Delta t}, \qquad k = 1, 2, \ldots, q. \tag{7.12}$$

It follows then from (7.6), (7.10), and (7.12) that

$$a_1 = \frac{2}{(\Delta t)^2}[x_1 - x_0 - v_0\,\Delta t], \tag{7.13}$$

$$a_2 = \frac{2}{(\Delta t)^2}[x_2 - 3x_1 + 2x_0 + v_0\,\Delta t], \tag{7.14}$$

$$a_k = \frac{2}{(\Delta t)^2}\left\{x_k - 3x_{k-1} + 2(-1)^k x_0 + 4\sum_{j=1}^{k-1}[(-1)^j x_{k-j}] + (-1)^k v_0 \Delta t\right\}, \qquad k \geq 3. \tag{7.15}$$

Note now that two-dimensional, three-dimensional, and, indeed, n-dimensional motion of a particle can be developed in the same spirit as that of Section 7.4 by the usual introduction of vectors. The vector formulas for velocity and acceleration would have as scalar components formulas of the types given in (7.6) and (7.10) through (7.15). However, we shall not have occasion to apply these formulas here, so we shall not carry through these generalizations but continue to consider motion only in a fixed direction.

7.6. Law of Motion

In Section 7.5 it was shown, among other things, how one could determine a particle's acceleration from a knowledge of its position at times $t_k = k\,\Delta t$, $k = 0, 1, \ldots, q$. In this and in the next section, we consider the problem of determining a particle's position from a knowledge of its acceleration at times $t_1, t_2, \ldots, t_q$. For simplicity, we consider only the case in which the particle's motion is in a fixed direction, say along an X axis.

At time $t_k = k\,\Delta t$, $k = 0, 1, \ldots, q$, let a particle of mass m be located at x_k. At time t_k, $k = 0, 1, \ldots, q - 1$, let the particle be acted upon by a force $F = F(t_{k-1}, x_{k-1}, v_{k-1})$. Then the motion of the particle is assumed to be governed by the generalized Newton's equation

$$m \cdot a(t_k) = F(t_{k-1}, x_{k-1}, v_{k-1}), \qquad k = 1, 2, \ldots, q. \tag{7.16}$$

The values of F can be given in either tabular form from experimental data or in the form of a mathematical expression, the former requiring only a greater memory capacity of the computer for its use.

There are several important facets of dynamical equation (7.16) which should be observed immediately. First note that, by (7.12), (7.16) is, in general, a first-order difference equation in v_i and recall that when solving difference equations we seek only discrete functions which are solutions. Next note that (7.13) can be solved easily for x_1, (7.14) can be solved easily for x_2, and (7.15) can be solved easily for each x_k, $k \geq 2$, so that (7.16) can be solved easily for each value of x_k, $k = 1, 2, \ldots, q$, whether F is linear or nonlinear in its variables. This latter observation, which will be illustrated more completely in Section 7.7, makes (7.16) exceptionally valuable computationally. Note finally that in applying (7.16) to physical problems, we shall tacitly assume the validity of Newton's first and third laws of equilibrium and action–reaction.

7.7. Damped Motion in a Nonlinear Force Field

In order to illustrate the application of (7.16), consider now a prototype problem of nonlinear mechanics. As shown in Figure 7.4, a particle of unit mass is constrained to move with its center Q on the X axis. A displacement of the particle such that the directed distance OQ is x is opposed by a field force of magnitude $\sin x$ and a viscous damping force of magnitude αv, $\alpha > 0$.

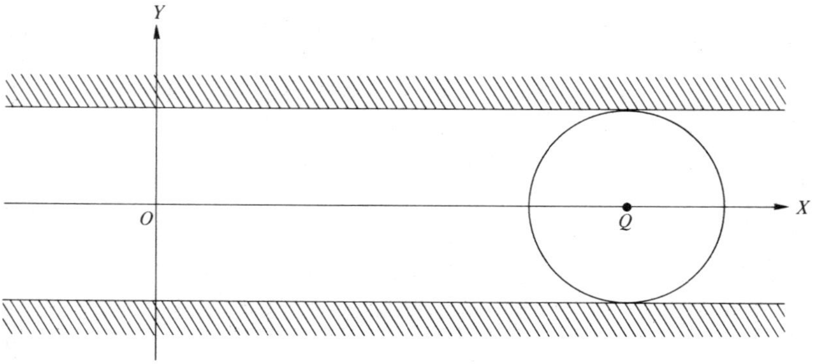

Figure 7.4

Then, from (7.16), the equation of motion is

$$a_i + \text{restoring force} = 0, \tag{7.17}$$

which, from the given data takes the particular form

$$a_i + \alpha v_{i-1} + \sin x_{i-1} = 0. \tag{7.18}$$

Substitution of (7.10) through (7.15) into (7.18) yields

$$\frac{2}{(\Delta t)^2}[x_1 - x_0 - v_0\,\Delta t] + \alpha v_0 + \sin x_0 = 0, \tag{7.19}$$

$$\frac{2}{(\Delta t)^2}[x_2 - 3x_1 + 2x_0 + v_0\,\Delta t] + \alpha\left\{\frac{2}{\Delta t}[x_1 - x_0] - v_0\right\} + \sin x_1 = 0, \tag{7.20}$$

$$\frac{2}{(\Delta t)^2}\left\{x_k - 3x_{k-1} + 2(-1)^k x_0 + 4\sum_{j=2}^{k-1}[(-1)^j x_{k-j}] + (-1)^k v_0\,\Delta t\right\} + \alpha\left\{\frac{2}{\Delta t}\left[x_{k-1} + (-1)^{k-1}x_0 + 2\sum_{j=1}^{k-2}(-1)^j x_{k-1-j}\right] + (-1)^{k-1}v_0\right\} + \sin x_{k-1} = 0, \qquad k \geq 3. \tag{7.21}$$

But now one sees explicitly the computational advantage of (7.16), for (7.19) can be solved easily for x_1 in terms of x_0 and v_0, (7.20) can be solved easily for x_2 in terms of x_1, x_0, and v_0, and (7.21) can be solved easily for x_k in terms of $x_{k-1}, x_{k-2}, \ldots, x_3, x_2, x_1, x_0$, and v_0 to yield the following equations for the motion of the particle:

$$(7.22)\ x_1 = x_0 + v_0\,\Delta t - \frac{(\Delta t)^2}{2}[\alpha v_0 + \sin x_0],$$

$$(7.23)\ x_2 = 3x_1 - 2x_0 - v_0\,\Delta t - \frac{(\Delta t)^2}{2}\left\{\alpha\left[\frac{2}{\Delta t}(x_1 - x_0) - v_0\right] + \sin x_1\right\},$$

$$(7.24)\ x_k = (3 - \alpha\,\Delta t)x_{k-1} + (-1)^{k-1}(2 - \alpha\,\Delta t)x_0 + (2\alpha\,\Delta t - 4)\sum_{j=2}^{k-1}[(-1)^j x_{k-j}] + (-1)^{k-1}\left(1 - \frac{\alpha\Delta t}{2}\right)v_0\,\Delta t - \frac{(\Delta t)^2}{2}\sin x_{k-1}, \qquad k \geq 3.$$

One now has the following theorem.

Theorem 7.1. For given x_0 and v_0, the solution of (7.18) exists, is unique, and is given constructively by (7.22) through (7.24).

Proof. Since the right-hand sides of (7.22) through (7.24) are always defined and real, the proof is an immediate consequence of the recursive structure of (7.22) through (7.24).

Note that the recursive structure of (7.22) through (7.24) makes (7.18) particularly suitable for solution on a digital computer. Note also that the simplicity of the proof of Theorem 7.1 contrasts remarkably with the amount of mathematical structure needed to deduce the existence and uniqueness of the corresponding initial-value problem for the differential analogue

$$\ddot{x} + \alpha\dot{x} + \sin x = 0$$

with x_0 and $\dot{x}_0$ given, for which, incidentally, no analytical method is known for actually constructing a solution.

Suppose now that the particle, as shown in Figure 7.4, is displaced to a point Q which has coordinate x_0, is held rigid (so that $v_0 = 0$), and is then released. We wish to trace the subsequent motion of the particle and consider the particular set of parameters $\alpha = 0.3$, $\Delta t = 0.01$, $x_0 = \pi/4$. For $T = 150$, the solution (7.22) through (7.24) was generated on the UNIVAC 1108 in 33 seconds. That portion of the solution between $T = 0$ and $T = 25$ is shown graphically in Figure 7.5 and exhibits strong damping and peak values of 0.754615, -0.480136, 0.297837, -0.185740, 0.116061, -0.072585, 0.045407, and -0.028421 which occur at times 0, 3.27, 6.48, 9.66, 12.84, 16.01, 19.19, and 22.36, respectively, and the time required for the particle to travel from one peak to the next decreased monotonically.

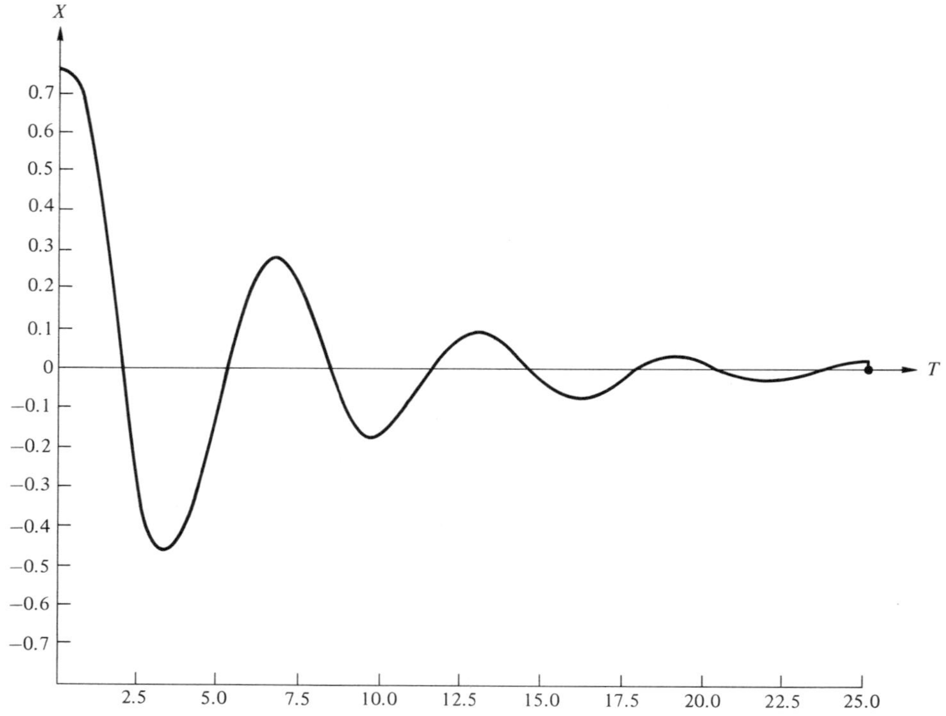

Figure 7.5

The remarkable resemblance between Figures 6.2 and 7.5 suggests, and correctly, that the particular problem considered above can be considered as a discrete form of the pendulum problem discussed in Section 6.9, and it is worth noting that though the two approaches yield almost identical results, the mathematics of the discrete model is completely arithmetic and by far the simpler.

Finally, it is worth noting that in the large variety of other examples run in which Δt and α were varied, it was found experimentally that difference equations (7.22) through (7.24) yielded instability only when one had $\Delta t > 2\alpha$. This result has now been established rigorously by Cryer.

7.8. Conservation of Energy

Let $t_k = k\,\Delta t, k = 0, 1, \ldots$. At time $t_i, i = 0, 1, 2, \ldots, n$ let a particle be located at point (x_i, y_i) which is on the straight line through $A(x_0, y_0)$ and $B(x_n, y_n)$, one possible arrangement of which is shown in Figure 7.6. Let $f(t_i), i = 0, 1, \ldots, n$ represent the component in the direction $\overrightarrow{AB}$ of a force $\vec{f}$ applied to the

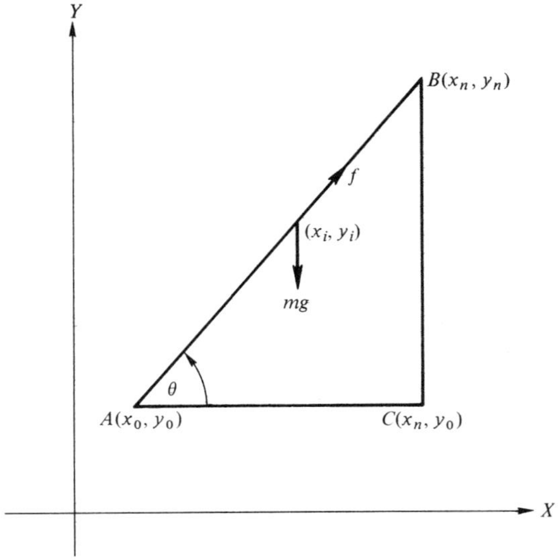

Figure 7.6

particle. Then the work W done by $\vec{f}$ in moving the particle in the fixed direction from A to B is defined to be

$$W = \sum_{i=1}^{n} f(t_{i-1})\,\Delta s_i, \tag{7.25}$$

where

$$\Delta s_i \equiv s_i - s_{i-1}$$

is the directed distance from (x_{i-1},y_{i-1}) to (x_i,y_i).

Now, let a_i and v_i represent the acceleration and velocity, respectively, at (x_i,y_i) in the direction $\overrightarrow{AB}$. Then, from (7.16) and (7.25) it follows that

$$\begin{aligned} W &= \sum_{i=1}^{n} f(t_{i-1})\,\Delta s_i \\ &= m \sum_{i=1}^{n} a_i\,\Delta s_i \\ &= m \sum_{i=1}^{n} \left[\left(\frac{v_i - v_{i-1}}{\Delta t}\right)(s_i - s_{i-1})\right] \\ &= m \sum_{i=1}^{n} \left[(v_1 - v_{i-1})\left(\frac{v_i + v_{i-1}}{2}\right)\right] \\ &= \frac{m}{2} \sum_{i=1}^{n} [v_i^2 - v_{i-1}^2]. \end{aligned}$$

Thus

$$W = \frac{mv_n^2}{2} - \frac{mv_0^2}{2}. \tag{7.26}$$

The quantity

$$K_i = \tfrac{1}{2}mv_i^2 \tag{7.27}$$

is defined to be the kinetic energy of the particle at time t_i. Equation (7.26) states that the work done in moving the particle from A to B is the difference of the kinetic energies at B and at A.

It is interesting, and somewhat surprising, to note now that the implicit formula (7.4) for velocity, and the explicit formula (7.12) for acceleration, neither of which is used commonly, have, with (7.16) and (7.25), resulted in the formula (7.27) for kinetic energy, which is identical to the usual one given in continuous Newtonian mechanics. A basic reason for this consistency lies in the identity

$$a_i\,\Delta s_i \equiv \frac{(v_i - v_{i-1})(v_i + v_{i-1})}{2},$$

for from it $a_i\,\Delta s_i$ can be summed easily over i, because

$$\sum_{i=1}^{n} [(v_i - v_{i-1})(v_i + v_{i+1})] \equiv \sum_{i=1}^{n} [v_i^2 - v_{i-1}^2]$$

is a telescopic sum. Formulas (7.26) and (7.27) do not follow from more usual definitions of velocity and acceleration because of the failure of certain key sums to telescope.

Neglecting friction, the force necessary to move a particle of mass m only along the vertical component of $\overrightarrow{AB}$ must be equal to the weight mg of the particle. Hence the amount of work done along the vertical component of the motion is

$$\begin{aligned} W &= mg(y_n - y_0) \\ &= mgy_n - mgy_0. \end{aligned}$$

The value

$$V_i = mgy_i \tag{7.28}$$

is called the *gravitational potential energy* of the particle at the point (x_i,y_i).

Now, as a particle of mass m is moved from A to B, neglecting friction one has

$$\bar{f}(t_{i-1}) = m \cdot a_i + mg \sin\theta, \tag{7.29}$$

where θ is the angle $\overrightarrow{AB}$ makes with the positive X axis, as shown in Figure 7.6, and $\bar{f}$ is the external force acting on the system. Then

$$\bar{f}(t_{i-1})\,\Delta s_i = ma_i\,\Delta s_i + mg(\sin\theta)\,\Delta s_i$$

$$= \frac{m}{2}(v_i^2 - v_{i-1}^2) + mg\,\Delta y_i.$$

Hence

$$\sum_{i=1}^{n} [\bar{f}(t_{i-1})\,\Delta s_i] = \frac{m}{2}v_n^2 - \frac{m}{2}v_0^2 + mgy_n - mgy_0$$

$$= K_n - K_0 + V_n - V_0.$$

In the special case when the external force acting on a particle is zero, so that $\bar{f} \equiv 0$, the last result implies that

$$K_n + V_n = K_0 + V_0, \tag{7.30}$$

which is called the *principle of conservation of energy*.

Thus the generalized Newton's equation (7.16) and our particular definitions of velocity, acceleration, and work have yielded a fundamental conservation principle in a completely arithmetic setting.

7.9. Momentum

Suppose two particles move with their centers on an X axis, as shown in Figure 7.7. Let the first particle have center A, mass m_A, and velocity $v(A)$, and let the second particle have center B, mass m_B, and velocity $v(B)$. We assume that the relative motion is such that the two bodies collide. During the collision, let the force of the first body on the second be F_A and that of the second on the first be F_B. These forces are equal and opposite but are nonconstant. Both are small at first, increase to a maximum, and then decrease to

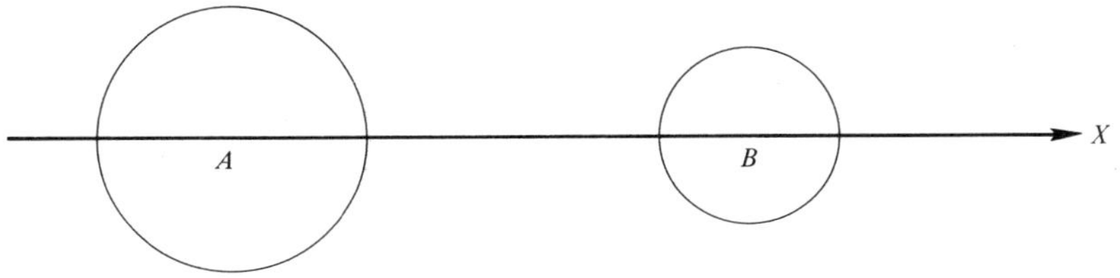

Figure 7.7

zero, at which time the bodies separate. Hence, at time t_i during the period of contact,

(7.31) $$F_A(t_{i-1}) = m_A a_i(A),$$

so

$$\sum_{i=1}^{n} [F_A(t_{i-1})\,\Delta t] = \sum_{i=1}^{n} m_A a_i(A)\,\Delta t$$

$$= m_A \sum_{i=1}^{n} [v_i(A) - v_{i-1}(A)],$$

(7.32) $$\sum_{i=1}^{n} [F_A(t_{i-1})\,\Delta t] = m_A[v_n(A) - v_0(A)].$$

Similarly,

(7.33) $$\sum_{i=1}^{n} [F_B(t_{i-1})\,\Delta t] = m_B[v_n(B) - v_0(B)].$$

The impulse I of a force F during the time of contact is defined to be

(7.34) $$I = \sum_{i=1}^{n} [F(t_{i-1})\,\Delta t]$$

and the momentum M at time t_i of a particle of mass m is defined by

(7.35) $$M = m \cdot v_i.$$

Then, since $F_A = -F_B$ at each time t_i, substitution into (7.32) and (7.33) implies that

$$m_A[v_n(A) - v_0(A)] = -m_B[v_n(B) - v_0(B)]$$

or, equivalently, that

(7.36) $$m_A v_0(A) + m_B v_0(B) = m_A v_n(A) + m_B v_n(B),$$

which is called the *conservation law of linear momentum.*

7.10. Angular Velocity and Acceleration

For completeness, and because of its importance in the study of the motion of large bodies, we conclude by developing the rotational analogues of the concepts, equations, and laws developed in Sections 7.5 through 7.9.

Let C be a circle whose center is O and whose radius is r (see Figure 7.8). Let a central angle θ be subtended from an arc of length s. Then the radian

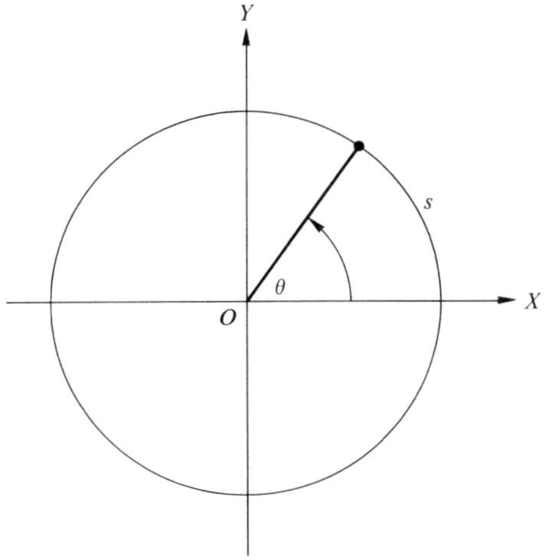

Figure 7.8

measure of θ is defined by

$$s = r\theta.$$

Next, let $t_k = k\,\Delta t, k = 0, 1, \ldots, n$. Let a particle be in motion on a circle of radius r and at time t_i let the angle subtended by the arc from $(r, 0)$ to the centroid of the particle be θ_i. Then the particle's *linear* velocity v_i at time t_i is defined implicitly by

$$\frac{v_i + v_{i-1}}{2} = \frac{s_i - s_{i-1}}{\Delta t} \quad \left(= \frac{r\theta_i - r\theta_{i-1}}{\Delta t}\right), \qquad i = 1, 2, \ldots, n,$$

and its *linear* acceleration a_i is defined by

$$a_i = \frac{v_i - v_{i-1}}{\Delta t}, \qquad i = 1, 2, \ldots, n.$$

By merely replacing x_i with s_i, all the previous results for motion in a fixed direction are now available to us. The particle's *angular* velocity $\omega(t_i) = \omega_i$, $i = 0, 1, \ldots, n$ and *angular* acceleration $\alpha(t_i) = \alpha_i$, $i = 1, 2, \ldots, n$, are defined simply by

$$v_i = r\omega_i, \qquad i = 0, 1, 2, \ldots, n \tag{7.37}$$

$$a_i = r\alpha_i, \qquad i = 1, 2, \ldots, n, \tag{7.38}$$

and it follows readily that

$$(7.39)\qquad \frac{\theta_i - \theta_{i-1}}{\Delta t} = \frac{\omega_i + \omega_{i-1}}{2}, \qquad i = 1, 2, \ldots, n,$$

$$(7.40)\qquad \alpha_i = \frac{\omega_i - \omega_{i-1}}{\Delta t}, \qquad i = 1, 2, \ldots, n,$$

and

$$(7.41)\qquad \omega_1 = \frac{2}{\Delta t}[\theta_1 - \theta_0] - \omega_0,$$

$$(7.42)\qquad \omega_n = \frac{2}{\Delta t}\left\{\theta_n + (-1)^n\theta_0 + 2\sum_{j=1}^{n-1}[(-1)^j\theta_{n-j}]\right\} + (-1)^n\omega_0, \qquad n \geq 2,$$

$$(7.43)\qquad \alpha_1 = \frac{2}{(\Delta t)^2}[\theta_1 - \theta_0 - \omega_0\,\Delta t],$$

$$(7.44)\qquad \alpha_2 = \frac{2}{(\Delta t)^2}[\theta_2 - 3\theta_1 + 2\theta_0 + \omega_0\,\Delta t],$$

$$(7.45)\qquad \alpha_n = \frac{2}{(\Delta t)^2}\left\{\theta_n - 3\theta_{n-1} + 2(-1)^n\theta_0 + 4\sum_{j=2}^{n-1}[(-1)^j\theta_{n-j}] + (-1)^n\omega_0\,\Delta t\right\}, \qquad n \geq 3.$$

Consider next a rigid system of particles, each of which is rotating about the origin O of the xy-coordinate system. The particles will be denoted by P_j, $j = 1, 2, \ldots, q$. Assume that each particle has the same angular velocity. For particle P_j, let m_j denote its mass and r_j its distance from O. If P_j's angular velocity is ω_i at time t_i, then the kinetic energy $k_{i,j}$ of P_j at time t_i is given by

$$(7.46)\qquad k_{i,j} = \tfrac{1}{2}m_j v_i^2 = \tfrac{1}{2}m_j r_j^2 \omega_i^2.$$

From (7.46), the kinetic energy $K(t_i) \equiv K_i$ of the system at time t_i is

$$(7.47)\qquad K_i = \sum_{j=1}^{q} \tfrac{1}{2}m_j r_j^2 \omega_i^2 = \frac{\omega_i^2}{2}\sum_{j=1}^{q} m_j r_j^2.$$

Defining the moment of inertia I of the system by

$$(7.48)\qquad I = \sum_{j=1}^{q} m_j r_j^2$$

implies that the kinetic energy K_i in (7.47) can be given by

$$(7.49)\qquad K_i = \tfrac{1}{2}I\omega_i^2.$$

At time t_i, let force $F_{i,j}$ act upon P_j. Then the moment $M_{i,j}$ of $F_{i,j}$ is

$$M_{i,j} = d_{i,j}F_{i,j}, \tag{7.50}$$

where $d_{i,j}$ is the perpendicular distance from O to the line of action of $F_{i,j}$. At time t_{i-1}, let P_j have traveled s_{i-1}, subtend angle θ_{i-1}, and be acted upon by force $F_{i-1,j}$, whose line of action is perpendicular to the radius to P_j. At time t_i, let P_i have traveled s_i and subtend angle θ_i. Then the work $W_{i,j}$ done in the time interval Δt on P_j is

$$W_{i,j} = (F_{i-1,j})\,\Delta s_i. \tag{7.51}$$

Hence

$$W_{i,j} = (F_{i-1,j})r_j\,\Delta\theta_i = M_{i-1,j}\,\Delta\theta_i. \tag{7.52}$$

The total work W_i done on the system during the period Δt is

$$W_i = \sum_{j=1}^{q} [M_{i-1,j}\,\Delta\theta_i], \tag{7.53}$$

while the total work W done on the system from the initial time t_0 until final time t_n is

$$W = \sum_{i=1}^{n}\sum_{j=1}^{q} [M_{i-1,j}\,\Delta\theta_i]. \tag{7.54}$$

Since the work done is equal to the change in kinetic energy, it follows from (7.49) and (7.53) that

$$\sum_{j=1}^{q} [M_{i-1,j}\,\Delta\theta_i] = \tfrac{1}{2}I(\omega_i^2 - \omega_{i-1}^2). \tag{7.55}$$

But (7.55) implies that

$$\sum_{j=1}^{q} M_{i-1,j} = I\frac{\omega_i - \omega_{i-1}}{\Delta t}\,\frac{\Delta t}{\Delta\theta_i}\,\frac{\omega_i + \omega_{i-1}}{2} = I\alpha_i,$$

so

$$\sum_{j=1}^{q} M_{i-1,j} = I\alpha_i, \tag{7.56}$$

which is the rotational analogue of (7.16).

The sum $\Sigma_{j=1}^{q} M_{i-1,j}$ is called the *torque* of the given system at time t_{i-1}.

The angular momentum L_i of a given system at time t_i is defined by

$$L_i = I\omega_i.$$

Hence

$$\sum_{j=1}^{q} M_{i-1,j} = \frac{L_i - L_{i-1}}{\Delta t},$$

so

$$\sum_{j=1}^{q} [M_{i-1,j}\,\Delta t] = L_i - L_{i-1},$$

(7.57)
$$\sum_{i=1}^{n}\sum_{j=1}^{q} [M_{i-1,j}\,\Delta t] = L_n - L_0.$$

Thus, if the torque of the system is zero at each time $t_0, t_1, \ldots, t_n$, then from (7.57),

(7.58)
$$L_n = L_0,$$

which is called the *law of conservation of angular momentum.*

7.11. Remark

Finally, it is rather interesting to note that in an effort to describe relativity in ultrasimplistic terms, Taylor and Wheeler have developed, in essence, the fundamentals of discrete *special relativistic mechanics.*

EXERCISES

1. Discuss the following free translation of Zeno's "Achilles and the tortoise" paradox. Show how the discrete time-and-motion model of Section 7.4 avoids such paradoxes.

   A fast runner and a slow tortoise are to have a race. Because of the runner's superior speed, the tortoise is allowed to begin the race at a positive distance d ahead of the runner (see the figure). Let the runner's

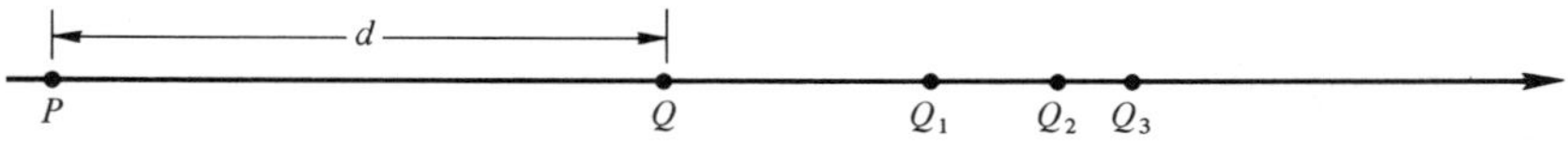

   initial point be P and that of the tortoise be Q. After the race has begun, the runner must reach the point Q, which takes time, during which the tortoise moves ahead to a new point Q_1. The runner must then reach the point Q_1, which takes time, during which the tortoise moves ahead to a new point Q_2. The runner must then reach the point Q_2, which takes time, during which the tortoise moves ahead to a new point Q_3, and so forth. Thus the runner must always reach a point where the tortoise has already been, from which it follows that the runner, no matter what his speed, can never overtake the tortoise.

2. A particle P is at rest on an X axis. During a 5-second time interval a force is exerted upon it and it travels to a new position. Let [0,5] be divided into five equal parts by the times $t_0 = 0$, $t_1 = 1$, $t_2 = 2$, $t_3 = 3$, $t_4 = 4$, $t_5 = 5$ and assume that, at time t_i, the directed distance x_i from the origin O to P, measured in centimeters, is given by

$$x(t_i) = x_i = 25 - (t_i)^2, \qquad i = 0, 1, 2, 3, 4, 5.$$

Then find P's location, velocity, and acceleration at t_1, t_2, t_3, t_4, t_5.

3. A 25-foot vertical cylinder which is closed at the bottom and open at the top is filled with water. A particle P is held at rest at the top of the liquid and then permitted to fall freely to the bottom. At intervals of 2 seconds the distances the particle has fallen are measured and recorded as follows:

$$0.04 \text{ ft at } t = 2 \text{ seconds},$$

$$1.44 \text{ ft at } t = 4 \text{ seconds},$$

$$4.84 \text{ ft at } t = 6 \text{ seconds},$$

$$10.24 \text{ ft at } t = 8 \text{ seconds},$$

$$17.64 \text{ ft at } t = 10 \text{ seconds}.$$

Find the velocity and acceleration of the particle at $t = 2, 4, 6, 8$.

4. Using the formulation in Section 7.7, describe the motion determined by each of the following:

(a) $a_i + 0.3v_{i-1} + \sin x_{i-1} = 0$, $x_0 = \pi/2$, $v_0 = 0$, $\Delta t = 0.01$

(b) $a_i + 0.3v_{i-1} + \sin x_{i-1} = 0$, $x_0 = \pi$, $v_0 = 0$, $\Delta t = 0.01$

(c) $a_i + 0.3v_{i-1} + \sin x_{i-1} = 0$, $x_0 = 99\pi/100$, $v_0 = 0$, $\Delta t = 0.01$

(d) $a_i + 0.3v_{i-1} + \sin x_{i-1} = 0$, $x_0 = -\pi/3$, $v_0 = 0$, $\Delta t = 0.01$

(e) $a_i + 0.6v_{i-1} + \sin x_{i-1} = 0$, $x_0 = \pi/4$, $v_0 = 0$, $\Delta t = 0.01$

(f) $a_i + 0.001v_{i-1} + \sin x_{i-1} = 0$, $x_0 = \pi/4$, $v_0 = 0$, $\Delta t = 0.01$

(g) $a_i + 0.001v_{i-1} + \sin x_{i-1} = 0$, $x_0 = \pi/4$, $v_0 = 0$, $\Delta t = 0.001$

5. Let AB, in Figure 7.6, be part of an S axis whose origin is at A and whose positive direction is that from A to B. Let the directed distance AB be 10 feet. Divide AB into n equal parts, each of length $10/n$, by means of the

points

$$s_0 = 0,\ s_1 = \frac{10}{n},\ s_2 = 2 \cdot \frac{10}{n},\ s_3 = 3 \cdot \frac{10}{n}, \ldots, s_n = 10.$$

A positive force f is then applied to move a particle from A to B. If f is given in pounds at each point $s_0, s_1, \ldots, s_n$, find the work done for each of the following cases:

(a) $n = 10, f(s_i) = 5$

(b) $n = 10, f(s_i) = s_i + 2$

(c) $n = 10, f(s_i) = s_i^2 + 1$

(d) $n = 100, f(s_i) = s_i + 1$

(e) $n = 100, f(s_i) = (1 + s_i)^{1/2}$

(f) $n = 20, f(0) = f(0.5) = f(1) = f(1.5) = 20.0$ lb,

$f(2) = f(2.5) = 20.2$ lb, $f(3) = f(3.5) = f(4) = f(4.5) = 19.9$ lb,

$f(5) = f(5.5) = f(6) = 18.7$ lb, $f(6.5) = 14.5$ lb, $f(7) = 12.2$ lb,

$f(7.5) = 6.3$ lb, $f(8) = 4.2$ lb, $f(8.5) = 2.1$ lb, $f(9) = 1.1$ lb,

$f(9.5) = 0.04$ lb, $f(10) = 0.00$ lb

6. A particle P is at rest at the point (1,0). During a 5-second time interval, a force is exerted upon it and it travels counterclockwise on the circle whose equation is $x^2 + y^2 = 1$ to a new point. Let $t_0 = 0$, $t_1 = 1$, $t_2 = 2$, $t_3 = 3$, $t_4 = 4$, $t_5 = 5$. Assume that at time t_i the central angle θ_i subtended by the path traced out by the particle is given by

$$\theta(t_i) = \theta_i = t_i^2 \pi/50.$$

Then find P's position, angular velocity, and angular acceleration at t_1, t_2, t_3, t_4, t_5.

7. By considering a vibrating string to consist of finitely many particles, each of which is moving in the vertical direction only, develop a discrete model of string vibration and the formulas needed to describe its motion.
8. Extend velocity formulas (7.6) through (7.10) and acceleration formulas (7.13) through (7.15) to the case of motion of a particle (a) in a plane, (b) in 3-space, and (c) in 4-space.

Chapter 8

INTERVAL ARITHMETIC

8.1. Introduction

Thus far, as in the discussion of stability in Chapter 6, only secondary interest has been shown in the fact that arithmetic operations are always subject to some roundoff procedure that is built into every digital computer. But because all the numerical methods developed thus far consist of finite sequences of arithmetic operations, it is necessary now for us to turn full attention to this matter. And because of the elementary nature of this book, we have selected a most elementary, but relatively efficient, approach to the problem, that of the interval arithmetic. For other approaches, see Conte, Hamming, Henrici (1), and Wilkinson.

As its name implies, interval arithmetic is an arithmetic of intervals. One is interested in such an arithmetic for the following kind of a reason. Suppose one has to compute

$$x + y,$$

where x and y are known only approximately. For example, if to three decimal places, x and y are 3.205 and 2.167, respectively, then all one really knows is that x is in the interval

$$3.2045 \le x \le 3.2055 \tag{8.1}$$

and that y is in the interval

$$2.1665 \le y \le 2.1675. \tag{8.2}$$

If one adds (8.1) and (8.2) termwise, which may be described as *adding the intervals*, then one concludes that

$$5.3710 \le x + y \le 5.3730, \tag{8.3}$$

and upper and lower bounds for the exact sum of the exact values of x and y have been obtained. Finally, if one takes as $x + y$ the midvalue of the interval (8.3), so that

$$x + y = 5.3720, \tag{8.4}$$

then one can conclude, because 5.3720 is the midvalue, that (8.4) is in error by at most ± 0.0010. Thus not only have we approximated the exact, but unknown, sum $x + y$, but we have bounds for the amount by which the approximation can be in error.

Since such approximations, with error bounds, would be of value for all arithmetic operations, not just addition, we shall develop a general interval arithmetic. Because the discussion will be highly set theoretic, it will be convenient to write an expression such as "A is a subset of B" in the usual set theoretic notation, "$A \subset B$." Also, as is customary, we allow every set to be a subset of itself, so that $A \subset B$ does not preclude A's being equal to B.

8.2. Interval Arithmetic

For $a < b$, the symbol $[a,b]$ often is used, as in this book, to represent the interval of numbers x that satisfy $a \leq x \leq b$. In interval arithmetic, however, it will be important to allow a to be equal to b, which leads to the following definition.

Definition 8.1. For $a \leq b$, the interval $[a;b]$ is the set of all real numbers x that satisfy $a \leq x \leq b$.

Example 1. $[-2;7]$ is the set of all real numbers x that satisfy $-2 \leq x \leq 7$.

Example 2. $[4;4]$ is the single number $x = 4$.

Definition 8.2. Let $I = [a;b]$, $J = [c;d]$. Then $I + J$ is the set of all real numbers $x + y$, where $x \in I$ and $y \in J$.

Example. Let $I = [1;2]$, $J = [5;9]$. Then, $1 \leq x \leq 2$, $5 \leq y \leq 9$. Hence termwise addition implies that $6 \leq x + y \leq 11$. Thus $[1;2] + [5;9] = [6;11]$.

Theorem 8.1. For two arbitrary intervals $[a;b]$ and $[c;d]$,

$$[a;b] + [c;d] \equiv [a + c;b + d].$$

Proof. If $x \in [a;b]$ and $y \in [c;d]$, then

$$a \leq x \leq b,$$
$$c \leq y \leq d.$$

Hence

$$a + c \leq x + y \leq b + d,$$

so that $x + y$ is in $[a + c;b + d]$. To complete the proof, we need only show that any number $z \in [a + c;b + d]$ can be written in the form $z = x + y$, where $x \in [a;b]$ and $y \in [c;d]$. But this follows easily because $z = x + y$ is a continuous function which is minimal at $x = a$, $y = c$, and is maximal at $x = b$, $y = d$, and because a continuous function takes on all intermediate values between its minimum and its maximum. Thus the theorem is proved.

Example 1. $[-4;7] + [5;9] = [1;16]$.

Example 2. $[-1;4] + [5;5] = [4;9]$.

Definition 8.3. Let $I = [a;b]$, $J = [c;d]$. Then $I - J$ is the set of all real numbers $x - y$ where $x \in I$ and $y \in J$.

Example. Let $I = [5;9]$, $J = [1;3]$. Then

$$5 \leq x \leq 9,$$
$$1 \leq y \leq 3$$

or

$$5 \leq x \leq 9,$$
$$-3 \leq -y \leq -1.$$

Hence, by termwise addition,

$$2 \leq x - y \leq 8,$$

so

$$[5;9] - [1;3] = [5;9] + [-3;-1] = [2;8].$$

Theorem 8.2. For two arbitrary intervals $[a;b]$ and $[c;d]$,

$$[a;b] - [c;d] \equiv [a;b] + [-d;-c] \equiv [a - d;b - c].$$

Proof. The proof is completely analogous to that of Theorem 8.1.

Example 1. $[2;5] - [4;3] = [2;5] + [-3;-4] = [-1;1]$.

Example 2. $[-1;0] - [-2;0] = [-1;0] + [0;2] = [-1;2]$.

Theorem 8.2 motivates the following consistent and convenient definition, which, in fact, enables one to view subtraction in the usual sense as the inverse of addition.

Definition 8.4. $-[a;b] \equiv [-b;-a]$.

Example. $-[-3;4] \equiv [-4;3]$.

Definition 8.5. If $I = [a;b]$ and $J = [c;d]$, then $I \cdot J$ is the set of all numbers $x \cdot y$ where $x \in I$ and $y \in J$.

Example 1. Consider $[3;5] \cdot [1;4]$. Then

$$3 \le x \le 5,$$

$$1 \le y \le 4,$$

from which it follows that

$$3 \le xy \le 20.$$

Thus

$$[3;5] \cdot [1;4] = [3;20].$$

Example 2. Consider $[-3;5] \cdot [1;4]$. Then

$$-3 \le x \le 5,$$

$$1 \le y \le 4.$$

But this time one has

$$-12 \le xy \le 20,$$

so that

$$[-3;5] \cdot [1;4] = [-12;20].$$

Example 3. Consider $[-3;-1] \cdot [1;4]$. Then

$$-3 \le x \le -1,$$

$$1 \le y \le 4,$$

from which one has

$$-12 \leq xy \leq -1.$$

Thus

$$[-3;-1]\cdot[1;4] = [-12;-1].$$

Example 4. Consider $[-3;-1]\cdot[-2;4]$. Then

$$-3 \leq x \leq -1,$$

$$-2 \leq y \leq 4.$$

Thus

$$-12 \leq xy \leq 6,$$

so that

$$[-3;-1]\cdot[1;4] = [-12;6].$$

From the above four examples, one notes that the product of two intervals $[\mathrm{a};\mathrm{b}]\cdot[\mathrm{c};\mathrm{d}]$ is an interval $[m;n]$, where $m =$ minimum (ac,ad,bc,bd) and $n =$ maximum (ac,ad,bc,bd). This leads to the following theorem.

Theorem 8.3. For two arbitrary intervals $[a;b]$ and $[c;d]$,

$$[a;\, b]\cdot[c;\, d] = [\min(ac,ad,bc,bd);\, \max(ac,ad,bc,bd)].$$

Proof. The proof is analogous to that of Theorem 8.1.

Example. $[0;7]\cdot[-1;1] = [\min(0,0,-7,7);\max(0,0,-7,7)] = [-7,7]$.

Definition 8.6. If 0 is not in $[c;d]$, then $[a;b] \div [c;d]$ is defined as the set of all real numbers x/y where $x \in [a;b]$ and $y \in [c;d]$.

Example. Consider $[1;3] \div [2;4]$. Then

$$1 \leq x \leq 3,$$

$$2 \leq y \leq 4.$$

Hence

$$1 \leq x \leq 3,$$

$$\frac{1}{4} \leq \frac{1}{y} \leq \frac{1}{2}.$$

Thus

$$\frac{1}{4} \leq \frac{x}{y} \leq \frac{3}{2}.$$

For the above example one sees that in Definition 8.6 one can consider x/y as the product $x(1/y)$, where $a \leq x \leq b$ and $1/d \leq 1/y \leq 1/c$. This converts division to multiplication and allows us to use Theorem 8.3. The result is as follows.

Theorem 8.4. For two intervals $[a;b]$ and $[c;d]$, where 0 is not in $[c;d]$,

$$\begin{aligned}[a;b] \div [c;d] &= [a;b] \cdot \left[\frac{1}{d};\frac{1}{c}\right] \\ &= \left[\min\left(\frac{a}{d},\frac{a}{c},\frac{b}{d},\frac{b}{c}\right);\max\left(\frac{a}{d},\frac{a}{c},\frac{b}{d},\frac{b}{c}\right)\right].\end{aligned}$$

Proof. The proof follows from Theorem 8.3.

Example 1. $[1;3] \div [2;4] = [1;3] \cdot [\frac{1}{4};\frac{1}{2}] = [\frac{1}{4};\frac{3}{2}]$.

Example 2. $[-1;2] \div [1;3] = [-1;2] \cdot [\frac{1}{3};1] = [-1;2]$.

It is then consistent with Definition 8.6 and Theorem 8.4 to add the following convenient definition.

Definition 8.7. If 0 is not in the interval $[c;d]$, then

$$\frac{1}{[c;d]} \equiv \left[\frac{1}{d};\frac{1}{c}\right].$$

Example 1. $\dfrac{1}{[-4;-1]} = \left[\dfrac{1}{-1};\dfrac{1}{-4}\right] = \left[-1;-\dfrac{1}{4}\right].$

Note now that for intervals of the form $[k;k]$, one has from Theorems 8.1 through 8.4 that

$$\begin{aligned}[a;a] + [b;b] &= [a+b;a+b] \\ [a;a] - [b;b] &= [a-b;a-b], \\ [a;a] \cdot [b;b] &= [ab;ab], \\ [a;a] \div [b;b] &= \left[\frac{a}{b};\frac{a}{b}\right], \qquad b \neq 0,\end{aligned}$$

from which it follows that under the biunique correspondence

$$[a;a] \leftrightarrow a,$$

the interval arithmetic includes the real arithmetic. We proceed then to consider a real number a to be identical with $[a;a]$.

Example 2. $7 + [1;2] = [7;7] + [1;2] = [8;9]$,

Example 3. $[-3;0] - 4 = [-3;0] - [4;4] = [-7;-4]$.

Example 4. $\frac{1}{2} \cdot [0;1] = [\frac{1}{2};\frac{1}{2}] \cdot [0;1] = [0;\frac{1}{2}]$.

Example 5. $[1;2] \div 2 = [1;2] \div [2;2] = [1;2] \cdot [\frac{1}{2};\frac{1}{2}] = [\frac{1}{2};1]$.

Note in particular that just as the numbers 0 and 1 are additive and multiplicative units, respectively, in the real number system, the corresponding intervals, that is,

$$0 \leftrightarrow [0;0],$$
$$1 \leftrightarrow [1;1]$$

are additive and multiplicative units, respectively, in the interval arithmetic, since for any interval $[a;b]$,

$$[a;b] + [0;0] = [0;0] + [a;b] = [a;b],$$
$$[a;b] \cdot [1;1] = [1;1] \cdot [a;b] = [a;b].$$

We consider next arithmetic combinations of three or more intervals. The naïve should observe that even for so simple a problem for integers as

$$17 + 12 + 14,$$

the actual arithmetic is not done in a single step; that is, the sum is not determined by adding the three numbers simultaneously. Indeed, what one does is to combine two numbers first and then add this result to the third. Thus

$$\begin{aligned} 17 + 12 + 14 &= (17 + 12) + 14 \\ &= 29 + 14 \\ &= 43. \end{aligned}$$

It is also a happy circumstance that if the arithmetic were performed as follows,

$$\begin{aligned} 17 + 12 + 14 &= 17 + (12 + 14) \\ &= 17 + 26 \\ &= 43, \end{aligned}$$

the answer is still the same. We extend these notions to interval arithmetic, and because Definitions 8.4 and 8.7 define subtraction in terms of addition, and division in terms of multiplication, we shall restrict the discussion to addition and multiplication only.

Definition 8.8. If $I = [a;b]$, $J = [c;d]$, $K = [e;f]$, then the sum $I + J + K$ is defined to be

$$I + J + K \equiv (I + J) + K$$

and the product $I \cdot J \cdot K$ is defined to be

$$I \cdot J \cdot K \equiv (I \cdot J) \cdot K.$$

Example. For $I = [-1;1]$, $J = [0;2]$, $K = [-3;-1]$,

$$\begin{aligned} [-1;1] + [0;2] + [-3;-1] &\equiv ([-1;1] + [0;2]) + [-3;-1] \\ &= [-1;3] + [-3;-1] \\ &= [-4;2], \end{aligned}$$

whereas

$$\begin{aligned} [-1;1] \cdot [0;2] \cdot [-3;-1] &= ([-1;1] \cdot [0;2]) \cdot [-3;-1] \\ &= [-2;2] \cdot [-3;-1] \\ &= [-6;6]. \end{aligned}$$

Theorem 8.5. If $I = [a;b]$, $J = [c;d]$, $K = [e;f]$, then

(8.5) $\quad I + J = J + I$ (commutative law of addition),

(8.6) $\quad I \cdot J = J \cdot I$ (commutative law of multiplication),

(8.7) $\quad (I + J) + K = I + (J + K)$ (associative law of addition),

(8.8) $\quad (I \cdot J) \cdot K = I \cdot (J \cdot K)$ (associative law of multiplication).

Proof. The proofs of (8.5) through (8.8) follow, essentially, from the commutative and associative laws of real numbers, as will be demonstrated by giving, in detail, the proofs of (8.5) and (8.7). For (8.5), one has

$$\begin{aligned} I + J &= [a;b] + [c;d] \\ &= [a + c;b + d] \\ &= [c + a;d + b] \\ &= [c;d] + [a;b] \\ &= J + I, \end{aligned}$$

whereas for (8.7) one has

$$\begin{aligned} (I + J) + K &= ([a;b] + [c;d]) + [e;f] \\ &= [a + c;b + d] + [e;f] \\ &= [a + c + e;b + d + f] \\ &= [a;b] + [c + e;d + f] \\ &= [a;b] + ([c;d] + [e;f]) \\ &= I + (J + K). \end{aligned}$$

The process of exponentiation will be treated as follows.

Definition 8.9. Let $I = [a;b]$. Then I^n, where n is a positive integer, is defined as the set of real numbers x^n, where $x \in [a;b]$.

Example 1. Consider $[2;3]^3$. Since

$$2 \leq x \leq 3,$$

one has

$$2^3 \leq x^3 \leq 3^3,$$

so

$$[2;3]^3 = [2^3;3^3].$$

Example 2. Consider $[-2;3]^3$. Since

$$-2 \leq x \leq 3,$$

one has

$$(-2)^3 \leq x^3 \leq 3^3,$$

so

$$[-2;3]^3 = [(-2)^3;3^3].$$

Example 3. Consider $[2;3]^4$. Since

$$2 \leq x \leq 3,$$

then

$$2^4 \leq x^4 \leq 3^4,$$

so

$$[2;3]^4 = [2^4;3^4].$$

Example 4. Consider $[-2;3]^4$. Since

$$-2 \leq x \leq 3,$$

we have

$$0 \leq x^4 \leq 3^4,$$

so

$$[-2;3]^4 = [0;3^4].$$

The above illustrative examples suggest the following theorem.

Theorem 8.6. If $I = [a;b]$ and n is a positive integer, then

$$I^n = \begin{cases} [a^n;b^n], & \text{if } a > 0, \\ [0;\max(a^n,b^n)], & \text{if 0 is in } I \text{ and } n \text{ is even,} \\ [a^n;b^n], & \text{if 0 is in } I \text{ and } n \text{ is odd,} \\ [b^n;a^n], & \text{if } b < 0 \text{ and } n \text{ is even,} \\ [a^n;b^n], & \text{if } b < 0 \text{ and } n \text{ is odd.} \end{cases}$$

Proof. The proof of the theorem is a direct consequence of the facts that a negative number to an odd power is negative and to an even power is positive, whereas a positive number to an odd power or an even power is positive.

It is worth noting that, in general

$$[a;b]^n \neq \underbrace{[a;b] \cdot [a;b] \cdots [a;b]}_{n \cdot \text{factors}}.$$

For example,

$$[-1;1]^2 = [0;1],$$

whereas

$$[-1;1] \cdot [-1;1] = [-1;1].$$

There is one final law of real numbers that plays a key role in arithmetic. It is the law that relates addition and multiplication and is called the distributive law because, in the form

$$a(b + c) = ab + ac, \tag{8.9}$$

it distributes the a by multiplication with each of b and c. Unfortunately, per se, (8.9) is *not valid* in interval arithmetic; that is, it is not always valid that

$$[a;b]([c;d] + [e;f]) \equiv [a;b]\cdot[c;d] + [a;b]\cdot[e;f].$$

The particular example

$$[-2;-1]([-3;-2] + [0;2])$$

is sufficient to illustrate the point, for

$$[-2;-1]([-3;-2] + [0;2]) = [-2;-1][-3;0] = [0;6],$$

whereas

$$[-2;-1][-3;-2] + [-2;-1][0;2] = [2;6] + [-4;0] = [-2;6].$$

It is interesting to note, however, that

$$[0;6] \subset [-2;6],$$

which, indeed, is the essence of the following theorem.

Theorem 8.7. If $I = [a;b]$, $J = [c;d]$, $K = [e;f]$, then

$$I(J + K) \subset (IJ + IK). \tag{8.10}$$

Proof. By direct computation,

$$\begin{aligned} I(J + K) &= [a;b]([c;d] + [e;f]) \\ &= [a;b][c + e;d + f] \\ &= [\min(ac + ae, ad + af, bc + be, bd + bf); \\ &\qquad \max(ac + ae, ad + af, bc + be, bd + bf)] \\ IJ + IK &= [a;b][c;d] + [a;b][e;f] \\ &= \min[(ac, ad, bc, bd); \max(ac, ad, bc, bd)] \\ &\quad + [\min(ae, af, be, bf); \max(ae, af, be, bf)] \\ &= [\min(ac, ad, bc, bd) + \min(ae, af, be, bf); \\ &\qquad \max(ac, ad, bc, bd) + \max(ae, af, be, bf)]. \end{aligned}$$

But, since

$$\min(ac, ad, bc, bd) + \min(ae, af, be, bf)$$
$$\leq \min(ac + ae, ad + af, bc + be, bd + bf),$$
$$\max(ac, ad, bc, bd) + \max(ae, af, be, bf)$$
$$\geq \max(ac + ae, ad + af, bc + be, bd + bf),$$

(8.10) is valid and the theorem is proved.

It should be noted that in later work we shall always wish to deal with the smallest interval possible, so *if one has the choice* of $I(J + K)$ and $(IJ + IK)$, (8.10) implies that one should always use the former.

8.3. General Application of Interval Arithmetic

The rationale behind the very extensive applications made of interval arithmetic (see, e.g., Moore), is founded on the following direct consequence of the definitions and theorems of Section 8.2. For exact values $x_1, x_2, \ldots, x_n$, suppose one wishes to determine $F(x_1, x_2, \ldots, x_n)$, where F is a given rational function. Suppose, however, $x_1, x_2, \ldots, x_n$ are known only approximately; that is, for intervals $I_1, I_2, \ldots, I_n$ one knows only that $x_1 \in I_1, x_2 \in I_2, \ldots, x_n \in I_n$, and that by the interval arithmetic one calculates $F(I_1, I_2, \ldots, I_n)$. Then

$$F(x_1, x_2, \ldots, x_n) \in F(I_1, I_2, \ldots, I_n).$$

We are thus led to the following convenient three-step algorithm for applying interval arithmetic.

Interval Arithmetic Algorithm.

Step 1. In performing a set of arithmetic operations in which only rounded numbers are available, replace the numbers by intervals that contain them. For example, suppose one wishes to determine x from

$$x = m^2 + 2n - c, \tag{8.11}$$

with $m = 0.75$, which is known to be correct to only two decimal places, with $n = 0.10056$, which is known to be correct to only five decimal places, and with $c = 0.00201$, which is known to be correct to only five decimal places. Then, in place of

$$m^2 + 2n - c, \tag{8.12}$$

we consider

$$[0.745;0.755]^2 + 2[0.100555;0.100565] - [0.002005;0.002015]. \tag{8.13}$$

Step 2. Combine the result of step 1 by the interval arithmetic to yield a single interval $[a;b]$. Thus, for example, (8.13) combines into

$$[0.754120;0.769150]. \tag{8.14}$$

Step 3. Take the midpoint of the interval $[a,b]$ generated by step 2 as an approximation to the desired value. This midvalue is $\bar{x} = (a + b)/2$. By virtue of $\bar{x}$'s being the midpoint, it is in error by at most $|(b - a)/2|$. Thus, for example, the midvalue $\bar{x}$ of (8.14) is

$$\bar{x} = 0.761635$$

and the error $|\bar{x} - x|$ is at most 0.007515; that is,

$$|\bar{x} - x| \leq \frac{0.769150 - 0.754120}{2} = 0.007515.$$

8.4. Application to Difference Equations

Because of the significance of difference equations in numerical analysis, we shall develop in this section algorithms for solving such equations with error bounds.

If $J = [a;b]$ is a given interval, it will be convenient to let

$$w = w(J) = \text{width } J = b - a. \tag{8.15}$$

For first-order difference equations, our algorithm will be formulated as follows.

Algorithm. Consider the first-order difference equation

$$y_{i+1} = F(x_i, y_i, \Delta x), \qquad i = 0, 1, \ldots, n - 1, \tag{8.16}$$

as defined in Section 6.3. Let $y_0 \in J_0$, $\Delta x \in K$, and $x_i \in I_i$, $i = 1, 2, \ldots, n$, where J_0, K, and I_i are intervals. Next, construct intervals J_{i+1}, $i = 0, 1, \ldots, n - 1$, by applying the interval arithmetic to

$$J_{i+1} = F(I_i, J_i, K), \qquad i = 0, 1, \ldots, n - 1, \tag{8.17}$$

and determine

$$w_i = \text{width } J_i, \qquad i = 0, 1, \ldots, n. \tag{8.18}$$

Finally, let

(8.19) $$\bar{y}_i = \text{midpoint } J_i, \qquad i = 0, 1, \ldots, n-1, n.$$

Then for each $i = 0, 1, \ldots, n$, the number $\bar{y}_i$ is to be taken as an approximation to the solution y_i of (8.16) for given y_0. Moreover,

$$|\bar{y}_i - y_i| \leq \frac{w_i}{2}, \qquad i = 0, 1, \ldots, n.$$

Example. Consider the difference equation

$$y_{i+1} = y_i + 2(\Delta x)x_i, \quad i = 0, 1, 2, \ldots,$$

and suppose one knows that

$$x_0 \in [0.325; 0.335] = I_0,$$

$$x_1 \in [0.665; 0.675] = I_1,$$

$$x_2 \in [0.95; 1.05] = I_2,$$

$$\Delta x \in [0.325; 0.335] = K,$$

$$y_0 \in [0.015; 0.025] = J_0.$$

Then consider

$$J_{i+1} = J_i + 2KI_i, \qquad i = 0, 1, 2, \ldots,$$

so

$$J_1 = J_0 + 2KI_0 = [0.226250; 0.249450],$$

$$J_2 = J_1 + 2KI_1 = [0.658500; 0.701700],$$

$$J_3 = J_2 + 2KI_2 = [1.276000; 1.405200].$$

Thus

$$w_0 = 0.010000,$$

$$w_1 = 0.023200,$$

$$w_2 = 0.043200,$$

$$w_3 = 0.129200,$$

whereas

$$\bar{y}_0 = 0.020, \qquad \bar{y}_1 = 0.237850, \qquad \bar{y}_2 = 0.680100, \qquad \bar{y}_3 = 1.340600,$$

which are in error by at most

$$|\bar{y}_0 - y_0| \le \frac{w_0}{2} = 0.005,$$

$$|\bar{y}_1 - y_1| \le \frac{w_1}{2} = 0.011600,$$

$$|\bar{y}_2 - y_2| \le \frac{w_2}{2} = 0.021600,$$

$$|\bar{y}_3 - y_3| \le \frac{w_3}{2} = 0.064600.$$

For second-order difference equations, one can formulate an algorithm completely analogous to the one above as follows.

Algorithm. Consider a second-order difference equation

(8.20) $$y_{i+1} = F(x_i, y_{i-1}, y_i, \Delta x), \qquad i = 1, 2, \ldots, n-1,$$

as defined in Section 4.6. Let $y_0 \in J_0$, $y_1 \in J_1$, $\Delta x \in K$, and $x_i \in I_i$, $i = 1, 2, \ldots, n$, where J_0, J_1, K, and I_i are intervals. Next, construct intervals J_{i+1}, $i = 1, 2, \ldots, n-1$, by applying interval arithmetic to

(8.21) $$J_{i+1} = F(I_i, J_{i-1}, J_i, K), \qquad i = 1, 2, \ldots, n-1,$$

and determine

(8.22) $$w_i = \text{width } J_i, \qquad i = 0, 1, \ldots, n.$$

Finally, let

(8.23) $$\bar{y}_i = \text{midpoint } J_i, \qquad i = 0, 1, \ldots, n.$$

Then for each $i = 0, 1, \ldots, n$, the number $\bar{y}_i$ is taken to be the approximate solution of (8.20), for given y_0 and y_1. Moreover,

$$|\bar{y}_i - y_i| \le \frac{w_i}{2}, \qquad i = 0, 1, \ldots, n.$$

Example. Suppose one is given the difference equation

(8.24) $$y_{i+1} = y_i + 2y_{i-1} - x_i\,\Delta x, \qquad i = 1, 2, 3,$$

and one knows that the data $x_1 = 0.4$, $x_2 = 0.6$, $x_3 = 0.8$, $y_0 = 4.2$,

$y_1 = 5.1$, $\Delta x = 0.2$ are correct to only one decimal place. Then

$$x_1 \in [0.35; 0.45] = I_1,$$
$$x_2 \in [0.55; 0.65] = I_2,$$
$$x_3 \in [0.75; 0.85] = I_3,$$
$$y_0 \in [4.15; 4.25] = J_0$$
$$y_1 \in [5.05; 5.15] = J_1,$$
$$\Delta x \in [0.15; 0.25] = K.$$

In place of (8.24) we consider

$$J_{i+1} = J_i + 2J_{i-1} - I_i \cdot K, \qquad i = 1, 2, 3.$$

Then

$$J_2 = J_1 + 2J_0 - I_1 \cdot K = [13.2375; 13.5975],$$
$$J_3 = J_2 + 2J_1 - I_2 \cdot K = [23.1750; 23.8150],$$
$$J_4 = J_3 + 2J_2 - I_3 \cdot K = [49.4375; 50.8975].$$

Thus

$$w_0 = 0.10, \quad w_1 = 0.10, \quad w_2 = 0.3600, \quad w_3 = 0.6400,$$
$$w_4 = 1.4600,$$
$$\bar{y}_0 = 4.20, \quad \bar{y}_1 = 5.10, \quad \bar{y}_2 = 13.4175, \quad \bar{y}_3 = 23.4950,$$
$$\bar{y}_4 = 50.1675,$$

and

$$|\bar{y}_0 - y_0| \leq 0.05,$$
$$|\bar{y}_1 - y_1| \leq 0.05,$$
$$|\bar{y}_2 - y_2| \leq 0.1800,$$
$$|\bar{y}_3 - y_3| \leq 0.3200,$$
$$|\bar{y}_4 - y_4| \leq 0.7300.$$

8.5. Remarks

Two observations are important in the practical application of interval arithmetic. First note that if one has to calculate an expression such as

(8.25) $$xy + xz,$$

where $x \in I$, $y \in J$, $z \in K$, then it is advantageous to rewrite (8.25) in the form

$$x(y + z). \tag{8.26}$$

For substitution of intervals into (8.25) yields

$$IJ + IK, \tag{8.27}$$

whereas substitution into (8.26) yields

$$I(J + K). \tag{8.28}$$

Since the error is determined by the width of the final interval, it follows from Theorem 8.7 that (8.28) has a width which is not greater than that of (8.27) and hence is to be considered the more desirable.

Finally, a very important rule must be observed when one is using interval arithmetic on a digital computer. The computation of intervals often requires rounding of numbers. To be assured always that the interval which results still contains the exact value one seeks to approximate, *one should always round the left number in an interval down and the right number up*, as illustrated in the following example.

Example. Suppose that in the process of doing interval arithmetic, a digital computer determines that the answer x lies in the interval $J = [0.11127; 0.21123]$. But suppose that the computer can carry only four decimal places so that the end points of J have to be rounded. If one uses any of the usual rules for rounding, then the computer would read out the interval $J_1 = [0.1113; 0.2112]$. Unfortunately, now, even though the solution x was in J, it need not be in J_1, for J is not contained in J_1. To be assured that $x \in J_1$, *one should always round the left-hand end point of J down, and the right-hand end point up*, so that J should have been rounded to $[0.1112; 0.2113]$.

EXERCISES

1. Reduce each of the following to a single interval:
   (a) $[1;4] + [-2;5]$
   (b) $[-3;-1] + [-7;11]$
   (c) $[2;34] + [-4;-4]$
   (d) $[1;7] - [1;9]$
   (e) $[-3;0] - [0;3]$
   (f) $[-8;-4] - [-6;-4]$
   (g) $[0;2] \cdot [1;8]$
   (h) $[-1;0] \cdot [-2;-1]$
   (i) $[-3;5] \cdot [1;3]$
   (j) $[-8;-7] \cdot [7;8]$
   (k) $[-7;-4] \cdot [-4;4]$
   (l) $[2;3] \div [3;5]$

(m) $[1;3] \div [-2;-1]$ (n) $[-2;2] \div [-2;-1]$

(o) $[-3;-1] \div [-2;-1]$ (p) $[1.243;2.687] - [6.6;6.6]$

(q) $[1.45;2.00] \div [3.19;3.20]$

2. Reduce each of the following to a single interval:

(a) $\dfrac{[1;3] - 4}{[2;3]}$ (b) $2 \cdot [1;4] - 3 \cdot [-1;0]$

(c) $\dfrac{2 \cdot [-3;-1] - 3 \cdot [1;2][2;3]}{[-4;-3] + 2 \cdot [5;7]}$ (d) $[2;3]^3$

(e) $[-2;3]^3$ (f) $[-3;-2]^3$

(g) $[2;3]^4$ (h) $[-2;3]^4$

(i) $[-3;-2]^4$ (j) $\dfrac{[1;3] - [-1;4] + 2[-2;3]^3}{3 + [-2;1]^2}$

3. Prove (8.6) and (8.8).
4. For n a positive integer, let $x_1, x_2, \ldots, x_n, y_1, y_2, \ldots, y_n$ be $2n$ real numbers. Prove that

$$\min(x_1,x_2,\ldots,x_n) + \min(y_1,y_2,\ldots,y_n) \\ \leq \min(x_1 + y_1, x_2 + y_2, \ldots, x_n + y_n),$$

$$\max(x_1,x_2,\ldots,x_n) + \max(y_1,y_2,\ldots,y_n) \\ \geq \max(x_1 + y_1, x_2 + y_2, \ldots, x_n + y_n).$$

5. In each of the following, the numbers have been rounded to the given number of decimal places. Use the interval arithmetic to compute each expression and give an error bound for each result:

(a) $71.22 + 64.35$ (b) $71.22 - 83.764$

(c) $71.22 + 64.3 - 83.7$ (d) $71.22 - (6.6)^2$

(e) $\dfrac{78.4 - 9.64}{40.0}$ (f) $\dfrac{78.4}{9.64 - 40.2}$

(g) $\dfrac{(89.5)(20.6) - (34.97)}{(2.785) + (7.6)(4.5)^2}$

6. Solve each of the following by interval arithmetic under the assumption that all given numbers have been rounded to the indicated number of decimal places:

(a) $y_{i+1} = y_i + 2(\Delta x)x_i, i = 0, 1, 2$

$$x_0 = 0.33 \qquad \Delta x = 0.33$$

$$x_1 = 0.66 \qquad y_0 = 1.12$$

$$x_2 = 0.99$$

(b) $y_{i+1} = \dfrac{y_i - 2(\Delta x)^2}{1 + (x_i)^2}, i = 0, 1, 2$

$$x_0 = 0.0 \qquad \Delta x = 0.33$$

$$x_1 = 0.3 \qquad y_0 = 2.1$$

$$x_2 = 0.7$$

(c) $y_{i+1} = \sin(x_i y_i), i = 0, 1, 2$

$$x_0 = 1.42137 \qquad \Delta x = 1.00000$$

$$x_1 = 2.42137 \qquad y_0 = 30.0$$

$$x_2 = 3.42137$$

7. Given that $x_1 = 0.4, x_2 = 0.6, x_3 = 0.8, y_0 = 4.2, y_1 = 5.1$, and $\Delta x = 0.2$ are correct to exactly one decimal place, apply the interval arithmetic to obtain approximate solutions and error bounds for each of the following difference equations:

   (a) $y_{i+1} = y_i + 2y_{i-1} - x_i \Delta x, i = 1, 2, 3$

   (b) $y_{i+1} = y_i - y_{i-1} - (\Delta x/x_i), i = 1, 2, 3$

   (c) $y_{i+1} = y_i^2 - y_{i-1}^2 + [\Delta x/(1 + x_i)], i = 1, 2, 3$

8. Repeat the computations of Exercise 7 under the assumption that x_1, x_2, x_3, and Δx are exact and compare the results with those of Exercise 7.

REFERENCES AND SOURCES FOR FURTHER READING

Abramowitz, M., and I. A. Stegun, *Handbook of Mathematical Functions*, National Bureau of Standards, Washington, D.C., 1965.

Ames, W. F., *Nonlinear Partial Differential Equations in Engineering*, Academic Press, New York, 1965.

Anselone, P. M., *Nonlinear Integral Equations*, University of Wisconsin Press, Madison, Wis., 1964.

Arden, B. W., *Digital Computing*, Addison-Wesley, Reading, Mass., 1963.

Bers, L., On Mildly Nonlinear Partial Differential Equations of Elliptic Type, *J. Res. Natl. Bur. Std.*, **51** (1953), pp. 229–236.

Bieberbach, L., *Theorie der Differentialgleichungen*, Dover, New York, 1944.

Booth, A. D., *Numerical Methods*, Butterworth, London, 1955.

Churchill, R. V., *Fourier Series and Boundary Value Problems*, McGraw-Hill, New York, 1941.

Coddington, E. A., and N. Levinson, *Theory of Ordinary Differential Equations*, McGraw-Hill, New York, 1955.

Collatz, L., (1) *The Numerical Treatment of Differential Equations*, Springer-Verlag, Berlin, 1959; (2) *Functional Analysis and Numerical Mathematics*, Academic Press, New York, 1966.

Conte, S. D., *Elementary Numerical Analysis*, McGraw-Hill, New York, 1965.

Courant, R., and D. Hilbert, *Methods of Mathematical Physics*, Vol. II, Wiley-Interscience, New York, 1962.

Cryer, C. W., Stability Analysis in Discrete Mechanics, *Tech. Rept. 67*, Department of Computer Sciences, University of Wisconsin, Madison, Wis., 1969.

Danby, J. M. A., *Fundamentals of Celestial Mechanics*, Macmillan, New York, 1962.

Davis, H. T., *Introduction to Nonlinear Differential and Integral Equations*, U.S. Atomic Energy Commission, Washington, D.C., 1960.

Davis, P. J., *Interpolation and Approximation*, Blaisdell, Waltham, Mass., 1963.

Davis, P. J., and P. Rabinowitz, *Numerical Integration*, Blaisdell, Waltham, Mass., 1967.

Fehlberg, E., Neue genauere Runge–Kutta-Formeln fur Differentialgleichungen *n*-ter Ordnung, *ZAMM*, **40** (1960), pp. 449–455.

Forsythe, G. E., Solving Linear Algebraic Equations Can Be Interesting, *Bull. AMS*, **59** (1953), pp. 299–329.

Forsythe, G. E., and W. R. Wasow, *Finite-Difference Methods for Partial Differential Equations*, Wiley, New York, 1960.

Fox, L., (1) *The Numerical Solution of Two-Point Boundary Problems in Ordinary Differential Equations*, Oxford University Press, New York, 1957; (2) *Numerical Solution of Ordinary and Partial Differential Equations*, Addison-Wesley, Reading, Mass., 1962; (3) *An Introduction to Numerical Linear Algebra*, Oxford University Press, New York, 1965.

Froberg, C. E., *Introduction to Numerical Analysis*, Addison-Wesley, Reading, Mass., 1965.

Galler, B. A., *The Language of Computers*, McGraw-Hill, New York, 1962.

Gangolli, R. A., and D. Ylvisaker, *Discrete Probability*, Harcourt, Brace & World, New York, 1967.

Garabedian, P. R., *Partial Differential Equations*, Wiley, New York, 1964.

Goldstein, H., *Classical Mechanics*, Addison-Wesley, Reading, Mass., 1959.

Golomb, M., and H. F. Weinberger, Optimal Approximation and Error Bounds, in *On Numerical Approximation*, University of Wisconsin Press, Madison, Wis., 1959, pp. 117–191.

Goodwin, E. T., *Modern Computing Methods*, Philosophical Library, New York, 1961.

Greenspan, D., (1) *Theory and Solution of Ordinary Differential Equations*, Macmillan, New York, 1960; (2) *Introduction to Partial Differential Equations*, McGraw-Hill, New York, 1961; (3) *Introductory Numerical Analysis of Elliptic Boundary Value Problems*, Harper & Row, New York, 1965; (4) *Numerical Solutions of Nonlinear Differential Equations*, Wiley, New York, 1966; (5) *Introduction to Calculus*, Harper & Row, New York, 1968; (6) *Lectures on the Numerical Solution of Linear, Singular and Nonlinear Differential Equations*, Prentice-Hall, Englewood Cliffs, N.J., 1969.

Greenspan, D., and S. V. Parter, Mildly Nonlinear Elliptic Partial Differential Equations and Their Numerical Solution, II, *Num. Math.*, **7** (1965), pp. 129–146.

Greville, T. N. E., *Theory and Applications of Spline Functions*, Academic Press, New York, 1969.

Hadley, G., *Linear Programming*, Addison-Wesley, Reading, Mass., 1962.

Hamming, R. W., *Numerical Methods for Scientists and Engineers*, McGraw-Hill, New York, 1962.

Heisenberg, W., *Physics and Philosophy*, Harper & Row, New York, 1958.

Hemmerle, W. J., *Statistical Computations on a Digital Computer*, Blaisdell, Waltham, Mass., 1967.

Henrici, P., (1) *Discrete Variable Methods in Ordinary Differential Equations*, Wiley, New York, 1962; (2) *Elements of Numerical Analysis*, Wiley, New York, 1964.

Heun, K., Neue Methode zur approximativen Integration der Differentialgleichungen einer unabhangigen Variable, *ZAMP*, **45** (1900), pp. 23–38.

Hildebrand, F. B., (1) *Advanced Calculus for Engineers*, Prentice-Hall, Englewood Cliffs, N.J., 1949; (2) *Introduction to Numerical Analysis*, McGraw-Hill, New York, 1953; (3) *Finite-Difference Equations and Simulations*, Prentice-Hall, Englewood Cliffs, N.J., 1968.

Hochstadt, H., *Special Functions of Mathematical Physics*, Holt, Rinehart and Winston, New York, 1961.

Householder, A. S., (1) *Principles of Numerical Analysis*, McGraw-Hill, New York, 1953; (2) *The Theory of Matrices in Numerical Analysis*, Blaisdell, Waltham, Mass., 1964.

Isaacson, E., and H. B. Keller, *Analysis of Numerical Methods*, Wiley, New York, 1966.

Janenko, N. N., *Difference Methods for Solutions of Problems in Mathematical Physics*, American Mathematical Society, Providence, R.I., 1967.

Kamke, E., *Differentialgleichungen*, Vols. I and II, Akademic-Verlag, Leipzig, 1959.

Kantorovich, L. V., and V. I. Krylov, *Approximate Methods of Higher Analysis*, Noordhoff, Groningen, 1958.

Keller, H. B., *Numerical Methods for Two-Point Boundary Value Problems*, Blaisdell, Waltham, Mass., 1968.

Kelly, L. G., *Handbook of Numerical Methods and Applications*, Addison-Wesley, Reading, Mass., 1967.

Kemeny, J. G., and J. L. Snell, *Mathematical Models in the Social Sciences*, Ginn & Company, Boston, 1962.

Kunz, K. S., *Numerical Analysis*, McGraw-Hill, New York, 1957.

Kutta, W., Beitrag zur naherungsweisen Integration totaler Differentialgleichungen, *ZAMP*, **46** (1901), pp. 435–453.

Lanczos, C., *Applied Analysis*, Prentice-Hall, Englewood Cliffs, N.J., 1956.

Langer, R. E., *On Numerical Approximation*, University of Wisconsin Press, Madison, Wis., 1959.

Leitmann, G., *Optimization Techniques*, Academic Press, New York, 1962.

Levy, H., and F. Lessman, *Finite Difference Equations*, Macmillan, New York, 1961.

Lorentz, G. G., *Approximation of Functions*, Holt, Rinehart and Winston, New York, 1966.

McLachlan, N. W., *Ordinary Non-linear Differential Equations in Engineering and Physical Sciences*, Oxford University Press, New York, 1950.

Milne, W. E., (1) *Numerical Calculus*, Princeton University Press, Princeton, N.J., 1950; (2) *Numerical Solution of Differential Equations*, Wiley, New York, 1953.

Mises, R. von, *Mathematical Theory of Compressible Fluid Flow*, Academic Press, New York, 1958.

Moore, R. E., *Interval Analysis*, Prentice-Hall, Englewood Cliffs, N.J., 1966.

Noble, B., *Numerical Methods*, Oliver & Boyd, London, 1964.

Ostrowski, A. M., *Solution of Equations and Systems of Equations*, 2nd ed., Academic Press, New York, 1966.

Petrovsky, I. G., *Partial Differential Equations*, Wiley-Interscience, New York, 1954.

Rall, L. B., (1) *Error in Digital Computers*, Vols. I and II, Wiley, New York, 1965; (2) *Computational Solution of Nonlinear Operator Equations*, Wiley, New York, 1969.

Ralston, A., *A First Course in Numerical Analysis*, McGraw-Hill, New York, 1965.

Ralston, A., and H. S. Wilf, *Mathematical Methods for Digital Computers*, Wiley, New York, 1960.

Rice, J. R., *The Approximation of Functions*, Addison-Wesley, Reading, Mass., 1964.

Richtmyer, R. D., and K. W. Morton, *Difference Methods for Initial-Value Problems*, 2nd ed., Wiley-Interscience, New York, 1967.

Rose, D. J., An Algorithm for Solving a Special Class of Tridiagonal Systems of Linear Equations, *Comm. ACM*, **12** (1969), pp. 234–236.

Sarafyan, D., Composite and Multi-step Runge–Kutta Formulas, *Tech. Rept. 18*, Louisiana State University, New Orleans, 1966.

Schechter, S., Iteration Methods for Nonlinear Problems, *Trans. AMS*, **104** (1962), pp. 179–189.

Schlichting, H., *Boundary Layer Theory*, McGraw-Hill, New York, 1960.

Smith, G. D., *Numerical Solution of Partial Differential Equations*, Oxford University Press, New York, 1965.

Soare, M., *Application of Finite Difference Equations to Shell Analysis*, Pergamon Press, Elmsford, N.Y., 1967.

Sokolnikoff, I. S., and R. M. Redheffer, *Mathematics of Physics and Modern Engineering*, McGraw-Hill, New York, 1958.

Southworth, R. W., and S. L. De Leeuw, *Digital Computation and Numerical Methods*, McGraw-Hill, New York, 1965.

Taylor, E. F., and J. A. Wheeler, *Spacetime Physics*, Freeman, San Francisco, 1966.

Todd, J., *Survey of Numerical Analysis*, McGraw-Hill, New York, 1962.

Tompkins, C. B., and W. L. Wilson, Jr., *Elementary Numerical Analysis*, Prentice-Hall, Englewood Cliffs, N.J., 1969.

Varga, R. S., *Matrix Iterative Analysis*, Prentice-Hall, Englewood Cliffs, N.J., 1962.

Wendroff, B., *Theoretical Numerical Analysis*, Academic Press, New York, 1966.

Weyl, H., *Space-Time-Matter*, Dover, New York, 1922.

Whitrow, G. J., *The Natural Philosophy of Time*, Harper & Row, New York, 1961.

Whittaker, E. T., *From Euclid to Eddington*, Dover, New York, 1958.

Whittaker, E. T., and G. N. Watson, *Modern Analysis*, Cambridge University Press, New York, 1952.

Wilkinson, J. H., *Rounding Errors in Algebraic Processes*, Prentice-Hall, Englewood Cliffs, N.J., 1963.

ANSWERS TO SELECTED EXERCISES

Chapter 1

3. (a) $L(x) = 4x,\ 0 \le x \le 2;\ 28x - 48,\ 2 \le x \le 4;\ 76x - 240,\ 4 \le x \le 6;\ 148x - 672,\ 6 \le x \le 8$
   (b) $L(x) = 0.01x,\ 0.0 \le x \le 0.1;\ 0.07x - 0.006,\ 0.1 \le x \le 0.2;\ 0.19x - 0.03,\ 0.2 \le x \le 0.3;\ 0.37x - 0.084,\ 0.3 \le x \le 0.4;\ 0.61x - 0.18,\ 0.4 \le x \le 0.5;\ 0.91x - 0.33, 0.5 \le x \le 0.6$
   (c) $L(x) = -11x + 8,\ 0 \le x \le 1;\ 5x - 8,\ 1 \le x \le 2;\ -2x + 6,\ 2 \le x \le 3;\ 0, 3 \le x \le 4$

4. (a) $P(x) = 6x^2 - 8x, 0 \le x \le 4; 18x^2 - 104x + 192, 4 \le x \le 8$
   (b) $P(x) = 0.3x^2 - 0.02x,\ 0.0 \le x \le 0.2;\ 0.9x^2 - 0.26x + 0.024,\ 0.2 \le x \le 0.4;\ 1.5x^2 - 0.74x + 0.12, 0.4 \le x \le 0.6.$
   (c) $P(x) = 8x^2 - 19x + 8, 0 \le x \le 2; x^2 - 7x + 12, 2 \le x \le 4$

5. (a) $P(x) = x,\ 0 \le x \le 2;\ (-\tfrac{3}{2})x^2 + (\tfrac{19}{2})x - 11,\ 2 \le x \le 4;\ 2x^2 - 21x + 55,\ 4 \le x \le 6$

   (b)
$$C(x) = \begin{cases} \dfrac{x(x-2)(x-3)}{2} - x(x-1)(x-3) + \dfrac{2x(x-1)(x-2)}{3}, 0 \le x \le 3 \\ -\dfrac{2(x-4)(x-5)(x-6)}{3} + \dfrac{3(x-3)(x-5)(x-6)}{2} \\ +\dfrac{(x-3)(x-4)(x-5)}{6}, 3 \le x \le 6 \end{cases}$$

   (c)
$$-\frac{x(x-2)(x-3)(x-4)(x-5)(x-6)}{120} + \frac{x(x-1)(x-3)(x-4)(x-5)(x-6)}{24}$$
$$-\frac{x(x-1)(x-2)(x-4)(x-5)(x-6)}{9} + \frac{x(x-1)(x-2)(x-3)(x-5)(x-6)}{16}$$
$$+\frac{x(x-1)(x-2)(x-3)(x-4)(x-5)}{720}$$

6. (a) $C(x) = -\frac{(x-0.1)(x-0.2)(x-0.3)}{0.006} + \frac{x(x-0.2)(x-0.3)}{0.001} - \frac{x(x-0.1)(x-0.2)}{0.006}$

(b) $Q(x) = \frac{(x-0.1)(x-0.2)(x-0.3)(x-0.4)}{0.0024} - \frac{x(x-0.2)(x-0.3)(x-0.4)}{0.0003}$

$+\frac{x(x-0.1)(x-0.2)(x-0.4)}{0.0006} + \frac{\alpha x(x-0.1)(x-0.2)(x-0.3)}{0.0024},$

where each of five distinct and arbitrary values of α will determine a unique quartic polynomial

10. $y = -0.62 + 1.52x$
12. $y = 1 - x^3$
13. $y = 1.4 \sin x - 1.6 \cos x + 2.25 \sin 2x - 4.08 \cos 2x$
18. $T_0(x) = 1,\ T_1(x) = x,\ T_2(x) = 2x^2 - 1,\ T_3(x) = 4x^3 - 3x,\ T_4(x) = 8x^4 - 8x^2 + 1$

Chapter 2

1. (a) 0.25694444
3. 0.74511935
5. (a) 0.25000000
7. 0.74683037
9. $y = (\frac{71}{21}) + (\frac{116}{35})x;\ \frac{175}{3}$
11. *Exact:* (a) 0, (b) 3, (c) 1, (d) 1, (e) −0.5, (f) e, (g) $-\pi$, (h) 1.38629436
    Two-point forward difference: (a) 0, (b) 3.30999960, (c) 1.09999985, (d) 1.90509945, (e) −0.45248866, (f) 2.78738500, (g) −3.12868880, (h) 1.38673186
12. *Three-point forward difference:* (a) 0, (b) 2.97999950, (c) 0.99999979, (d) 0.76859936, (e) −0.49514128, (f) 2.71592890, (g) −3.16720780, (h) 1.38622522
13. $E = \frac{(\Delta x)^3}{8} f^{(iv)}(\xi_1) - (\Delta x)^3 f^{(iv)}(\xi_2) + \frac{9}{8}(\Delta x)^3 f^{(iv)}(\xi_3)$, where $x_i < \xi_1 < x_{i+1} < \xi_2 < x_{i+2} < \xi_3 < x_{i+3}$
14. $E = \frac{(\Delta x)^4}{30} f^{(v)}(\xi_1) - \frac{4}{5}(\Delta x)^4 f^{(v)}(\xi_2) + \frac{27}{10}(\Delta x)^4 f^{(v)}(\xi_3) - \frac{32}{15}(\Delta x)^4 f^{(v)}(\xi_4)$,
    where $x_i < \xi_1 < x_{i+1} < \xi_2 < x_{i+2} < \xi_3 < x_{i+3} < \xi_4 < x_{i+4}$
16. *Exact:* (a) 0, (b) 4, (c) −1, (d) 7, (e) −0.5, (f) e, (g) $-\pi$, (h) 5.43656366
    Two-point backward difference: (a) 0, (b) 3.70987500, (c) −1.20000005, (d) 6.03325400, (e) −0.55248615, (f) 2.58678730, (g) −3.14157680, (h) 4.70373870
    Three-point backward difference: (a) 0, (b) 3.98074990, (c) −0.99999994, (d) 6.84947710, (e) −0.49521626, (f) 2.70987000, (g) −3.14159220, (h) 5.29847240
18. *Exact:* (a) 0, (b) 12, (c) 2, (d) 42, (e) 1, (f) e, (g) 0, (h) 16.30969098
    Three-point forward difference: (a) 0, (b) 13.23499690, (c) 1.99999810, (d) 53.80650200, (e) 0.85305237, (f) 3.00667290, (g) 0.00128057, (h) 23.20083400
19. (a) 0, (b) 10.83499790, (c) 2.00000410, (d) 32.64892100, (e) 1.14539798, (f) 2.46165690, (g) −0.03093737, (h) 11.89467620
20. (a) 0, (b) 12.00498360, (c) 2.00000110, (d) 42.17505500, (e) 0.99997493, (f) 2.72053780, (g) 0.02974412, (h) 16.48288400

Chapter 3

3. (a) $L = \begin{pmatrix} -2 & 0 & 0 & 0 \\ 1 & -\frac{3}{2} & 0 & 0 \\ 0 & 1 & -\frac{4}{3} & 0 \\ 0 & 0 & 1 & -\frac{5}{4} \end{pmatrix}, U = \begin{pmatrix} 1 & -\frac{1}{2} & 0 & 0 \\ 0 & 1 & -\frac{2}{3} & 0 \\ 0 & 0 & 1 & -\frac{3}{4} \\ 0 & 0 & 0 & 1 \end{pmatrix}$

(b) $L = \begin{pmatrix} -3 & 0 & 0 & 0 \\ 2 & -\frac{7}{3} & 0 & 0 \\ 0 & 2 & -\frac{15}{7} & 0 \\ 0 & 0 & 2 & -\frac{31}{15} \end{pmatrix}, U = \begin{pmatrix} 1 & -\frac{1}{3} & 0 & 0 \\ 0 & 1 & -\frac{3}{7} & 0 \\ 0 & 0 & 1 & -\frac{7}{15} \\ 0 & 0 & 0 & 1 \end{pmatrix}$

(c) $L = \begin{pmatrix} 10 & 0 & 0 & 0 \\ 4 & \frac{56}{5} & 0 & 0 \\ 0 & 4 & \frac{155}{14} & 0 \\ 0 & 0 & 4 & \frac{1718}{155} \end{pmatrix}, U = \begin{pmatrix} 1 & -\frac{3}{10} & 0 & 0 \\ 0 & 1 & -\frac{15}{56} & 0 \\ 0 & 0 & 1 & -\frac{42}{155} \\ 0 & 0 & 0 & 1 \end{pmatrix}$

(d) $L = \begin{pmatrix} 5 & 0 & 0 & 0 & 0 \\ -2 & \frac{21}{5} & 0 & 0 & 0 \\ 0 & -2 & \frac{85}{21} & 0 & 0 \\ 0 & 0 & -2 & \frac{341}{85} & 0 \\ 0 & 0 & 0 & -2 & \frac{1365}{341} \end{pmatrix}, U = \begin{pmatrix} 1 & -\frac{2}{5} & 0 & 0 & 0 \\ 0 & 1 & -\frac{10}{21} & 0 & 0 \\ 0 & 0 & 1 & -\frac{42}{85} & 0 \\ 0 & 0 & 0 & 1 & -\frac{170}{341} \\ 0 & 0 & 0 & 0 & 1 \end{pmatrix}$

5. (a) $x_1 = 1,\ x_2 = 0,\ x_3 = 0,\ x_4 = 0$, (b) $x_1 = 1,\ x_2 = x_3 = x_4 = 0,\ x_5 = -1$, (d) $x_1 = 1, x_2 = x_3 = x_4 = 0$
6. (a) $x_1^{(6)} = 3.000,\ x_2^{(6)} = 0.000$, (b) $x_1^{(6)} = 0.999,\ x_2^{(6)} = -1.000$, (c) $x_1^{(6)} = 0.288$, $x_2^{(6)} = -0.182$
7. (a) $x_1^{(6)} = 2.093,\ x_2^{(6)} = 0.205$, (b) $x_1^{(6)} = -2.246,\ x_2^{(6)} = -3.264$, (c) $x_1^{(6)} = 0.180$, $x_2^{(6)} = -0.125$.
8. (a) $x_1^{(5)} = 0.62, x_2^{(5)} = -0.15, x_3^{(5)} = -1.23$, (b) $x_1^{(5)} = x_2^{(5)} = x_3^{(5)} = 0.00$
9. (a) $x_1^{(5)} = 1.62, x_2^{(5)} = -0.42, x_3^{(5)} = 0.28$, (b) $x_1^{(5)} = x_2^{(5)} = x_3^{(5)} = 0.00$
10. $x_1^{(3)} = 1.000, x_2^{(3)} = -1.246, x_3^{(3)} = 0.153, x_4^{(3)} = 2.061, x_5^{(3)} = -1.955, x_6^{(3)} = 0.025$, $x_7^{(3)} = -2.928$

Chapter 4

1. (a) solution, (b) solution, (c) not a solution
3. (a) $y = c_1e^{-5x} + c_2e^{-3x}$, (b) $y = c_1e^{-5x} + c_2e^{3x}$, (c) $y = c_1e^{3x} + c_2xe^{3x}$, (d) $y = c_1e^{-3x} + c_2xe^{-3x}$, (e) $y = c_1e^{2x} + c_2e^{x}$, (f) $y = c_1e^{2x}\cos 3x + c_2e^{2x}\sin 3x$, (g) $y = c_1 + c_2e^{(-3/2)x}$ (h) $y = c_1\cos 5x + c_2\sin 5x$, (i) $y = c_1e^{(-1/8)x}\cos(\frac{\sqrt{15}}{8})x + c_2e^{(-1/8)x}\sin(\frac{\sqrt{15}}{8})x$, (j) $y = c_1 + c_2x$
4. (a) $y = c_1e^{3x} + c_2e^{x} + \frac{1}{3}$, (b) $y = c_1e^{-2x} + c_2e^{x} + x^2$, (c) $y = c_1\cos x + c_2\sin x - 5x + x^3$, (d) $y = c_1e^{3x} + c_2xe^{3x} + e^{2x}$, (f) $y = c_1\cos x + c_2\sin x + x^3 - 6x - \frac{1}{2}xe^{x} - (x/2)\sin x$

6. (c) $y = (c_1/x)\cos x + (c_2/x)\sin x$
   (d) $y = c_1 \cos (x)^{1/2} + c_2 \sin (x)^{1/2}$
   (e) $y = c_1 e^x + c_2 \sum_0^\infty \dfrac{2^n x^{n+(1/2)}}{(2n+1)(2n-1)(2n-3)\cdots 5\cdot 3\cdot 1}$
   (g) $y = (c_1\sqrt{x} + c_2 x) \sum_{n=0}^\infty (-1)^{n+1} x^n$
   (h) $y = c_1 e^{-x} + c_2(1-x)$
   (i) $y = c_1 x^{1/2} e^x + c_2 x^{1/2} e^x \log|x|$
   (j) $y = (c_1 + c_2 \log|x|) \sum_0^\infty \dfrac{(-1)^{n+1} x^{2n+1}}{2^{2n} n!}$
10. (a) $y_0 = 0$, $y_1 = 1$, $y_2 = 1$, $y_3 = 3$, $y_4 = 3.33$, $y_5 = 4.90$, $y_6 = 5.68$, $y_7 = 6.86$, $y_8 = 7.83$, $y_9 = 8.88$, $y_{10} = 9.88$
   (b) $y_0 = 1$, $y_1 = 1$, $y_2 = 2$, $y_3 = 2.25$, $y_4 = 2.69$, $y_5 = 3.00$, $y_6 = 3.29$, $y_7 = 3.54$, $y_8 = 3.76$, $y_9 = 3.95$, $y_{10} = 4.13$
   (c) $y_0 = 1$, $y_1 = 0$, $y_2 = -1$, $y_3 = -2$, $y_4 = -3$, $y_5 = -4$, $y_6 = -5$, $y_7 = -6$, $y_8 = -7$, $y_9 = -8$, $y_{10} = -9$
11. (a) $y_0 = 0$, $y_1 = 0.330$, $y_2 = 0.600$, $y_3 = 0.820$, $y_4 = 1.000$
   (b) $y_0 = 1.000$, $y_1 = 0.655$, $y_2 = 0.382$, $y_3 = 0.167$, $y_4 = 0.000$

Chapter 5

2. (a) *Exact:* $y = x$; *numerical:* $y_0 = 0.0$, $y_1 = 0.2$, $y_2 = 0.4$, $y_3 = 0.6$, $y_4 = 0.8$, $y_5 = 1.0$
   (b) *Exact:* $y = \frac{5}{6}x + \frac{1}{6}x^3$; *numerical:* $y_0 = 0.000$, $y_1 = 0.168$, $y_2 = 0.344$, $y_3 = 0.536$, $y_4 = 0.752$, $y_5 = 1.000$
   (c) *Exact:* $y = 0.421e^x - 0.421e^{-x}$; *numerical:* $y_0 = 0.0000$, $y_1 = 0.1714$, $y_2 = 0.3497$, $y_3 = 0.5419$, $y_4 = 0.7558$, $y_5 = 1.0000$
   (f) *Exact:* $y = 0.421e^x - 0.421e^{-x} - x$; *numerical:* $y_0 = 0.0000$, $y_1 = 0.1428$, $y_2 = 0.2993$, $y_3 = 0.4838$, $y_4 = 0.7117$, $y_5 = 1.0000$
3. (a) $y_0 = 0.0000$, $y_1 = 0.1481$, $y_2 = 0.3423$, $y_3 = 0.6429$, $y_4 = 1.1436$, $y_5 = 2.0000$
   (b) $y_0 = -1.0000$, $y_1 = -0.6197$, $y_2 = -0.2703$, $y_3 = 0.0569$, $y_4 = 0.3720$, $y_5 = 0.6837$, $y_6 = 1.0000$
   (c) $y_0 = 0.0000$, $y_1 = -0.0028$, $y_2 = -0.0058$, $y_3 = -0.0097$, $y_4 = -0.0165$, $y_5 = -0.0336$, $y_6 = -0.0924$, $y_7 = -0.3211$, $y_8 = -1.0000$
4. (a) $y_0 = 0.0000$, $y_1 = -0.7302$, $y_2 = -0.9787$, $y_3 = -0.8513$, $y_4 = -0.2971$, $y_5 = 1.0000$
   (b) $y_0 = 0.0000$, $y_1 = 0.2871$, $y_2 = 0.5220$, $y_3 = 0.7141$, $y_4 = 0.8714$, $y_5 = 1.0000$

Chapter 6

2. (a) $y = \frac{1}{2}x^2 + 1$, (b) $y = 1 + x - 2e^x$, (d) $y = \exp(x^3/3)$
4. (a) $y_0 = 1.00$, $y_1 = 1.00$, $y_2 = 1.04$, $y_3 = 1.12$, $y_4 = 1.24$, $y_5 = 1.40$
   (b) $y_0 = -1.000$, $y_1 = -1.200$, $y_2 = -1.480$, $y_3 = -1.856$, $y_4 = -2.347$, $y_5 = -2.977$

(c) $y_0 = 4.000$, $y_1 = 4.000$, $y_2 = 4.080$, $y_3 = 4.246$, $y_4 = 4.516$, $y_5 = 4.919$
(d) $y_0 = 1.000$, $y_1 = 1.200$, $y_2 = 1.488$, $y_3 = 1.931$, $y_4 = 2.676$, $y_5 = 4.103$
(e) $y_0 = 1.000$, $y_1 = 0.600$, $y_2 = 0.373$, $y_3 = 0.222$, $y_4 = 0.096$, $y_5 = -0.041$

8. (a) $y_0 = 1.00$, $y_1 = 1.02$, $y_2 = 1.08$, $y_3 = 1.18$, $y_4 = 1.32$, $y_5 = 1.50$
(b) $y_0 = -1.000$, $y_1 = -1.240$, $y_2 = -1.577$, $y_3 = -2.032$, $y_4 = -2.631$, $y_5 = -3.406$
(c) $y_0 = 4.000$, $y_1 = 4.040$, $y_2 = 4.156$, $y_3 = 4.393$, $y_4 = 4.751$, $y_5 = 5.290$
(d) $y_0 = 1.000$, $y_1 = 1.244$, $y_2 = 1.640$, $y_3 = 2.383$, $y_4 = 4.190$, $y_5 = 11.879$
(e) $y_0 = 1.0000$, $y_1 = 0.6835$, $y_2 = 0.4781$, $y_3 = 0.3137$, $y_4 = 0.1547$, $y_5 = -0.0267$

12. *Heun's formula:* $y_0 = 1.00000$, $y_1 = 0.90961$, $y_2 = 0.83745$, $y_3 = 0.78162$, $y_4 = 0.74062$, $y_5 = 0.71303$, $y_6 = 0.69759$, $y_7 = 0.69314$, $y_8 = 0.69863$, $y_9 = 0.71311$, $y_{10} = 0.73573$
Kutta's formula: $y_0 = 1.00000$, $y_1 = 0.90892$, $y_2 = 0.83617$, $y_3 = 0.78000$, $y_4 = 0.73881$, $y_5 = 0.71118$, $y_6 = 0.69579$, $y_7 = 0.69147$, $y_8 = 0.69716$, $y_9 = 0.71191$, $y_{10} = 0.73576$

13. (a) $y_0 = 0.78539816$, $y_1 = 0.78536285$, $y_2 = 0.78525703$, $y_3 = 0.78508095$, $y_4 = 0.78483480$, $y_5 = 0.78451883$
(b) $y_0 = 1.57079633$, $y_1 = 1.57074633$, $y_2 = 1.57059634$, $y_3 = 1.57034634$, $y_4 = 1.56999634$, $y_5 = 1.56954636$
(c) $y_0 = y_1 = y_2 = y_3 = y_4 = y_5 = 3.14159270$

15. (a) stable, (b) stable, (c) stable

16. (a) stable, (b) stable, (c) unstable, (d) stable, (e) stable

Chapter 7

2.

i	t_i	x_i	v_i	a_i
1	1	24	−2	−2
2	2	21	−4	−2
3	3	16	−6	−2
4	4	9	−8	−2
5	5	0	−10	−2

3.

i	t_i	x_i	v_i	a_i
1	2	0.04	0.04	0.02
2	4	1.44	1.36	0.66
3	6	4.84	2.04	0.34
4	8	10.24	3.36	0.66

4. (a) Up to $T = 25$, peak x values of 1.5707964, −0.8896165, 0.5401099, −0.3344006, 0.2085199, −0.1303190, 0.0815091, −0.0510177 at respective times 0.00, 3.58, 6.87, 10.09, 13.28, 16.45, 19.61, 22.79
(b) Up to $T = 25$, peak x values of 3.1415927, −1.2282256, 0.7245731, −0.4445861, 0.2762522 at respective times 0.00, 12.61, 16.01, 19.27, 22.47

(c) Up to $T = 25$, peak x values of 3.1101768, −1.2281176, 0.7245193, −0.4445879, 0.2762994, −0.1724300, 0.1078173 at respective times 0.00, 7.94, 11.35, 14.60, 17.80, 20.98, 24.15

(d) Up to $T = 25$, peak x values of −1.0471976, 0.6281354, −0.3872378, 0.2409290, −0.1503590, 0.0940227, −0.0588625, 0.0368330 at respective times 0.00, 3.34, 6.58, 9.77, 12.95, 16.12, 19.29, 22.45

(f) unstable

5. (a) 50.000000, (b) 65.000000, (c) 295.000000, (d) 59.499988, (e) 23.539120, (f) 148.269990

6.

i	t_i	θ_i	ω_i	α_i
1	1	0.06283185	0.12566371	0.12566371
2	2	0.25132741	0.25132741	0.12566371
3	3	0.56548667	0.37699111	0.12566372
4	4	1.00530963	0.50265479	0.12566370
5	5	1.57079630	0.62831855	0.12566376

Chapter 8

1. (a) [−1;9], (b) [−10;10], (c) [−2;30], (d) [−8;6], (e) [−6;0], (f) [−4;2], (g) [0;16], (h) [0;2], (i) [−9;15], (j) [−64;−49], (k) [−28;28], (l) [$\frac{2}{5}$;1], (m) [−3;−$\frac{1}{2}$], (n) [−2;2], (o) [$\frac{1}{2}$;3], (p) [−5.357;−3.913], (q) [0.45;0.63]
2. (b) [2;11], (c) [−4;−$\frac{8}{11}$], (d) [8;27], (e) [−8;27], (f) [−27;−8], (g) [16;81], (h) [0;81], (i) [16;81], (j) [−$\frac{19}{3}$;58/3]
5. (a) 135.570 ± 0.010, (b) −12.5440 ± 0.0055, (c) 51.820 ± 0.105, (d) 27.6575 ±0.6650, (e) 1.71901 ± 0.00352, (f) −2.565456 ± 0.006253, (g) 11.55093 ±0.361884
6. (a) 1.33785 ± 0.0116, 1.7735 ± 0.0215, 2.42695 ± 0.0347
   (b) 1.87987375 ± 0.05887625, 1.5240982 ± 0.1017842, 0.88055105 ± 0.11391815
7. (a) 4.2 ± 0.05, 5.1 ± 0.05, 13.4175 ± 0.1800, 23.4950 ± 0.3200, 50.1675 ± 0.7300
   (b) 4.2 ± 0.05, 5.1 ± 0.05, 0.3762 ± 0.2905, −5.06645 ± 0.45245, −5.697555 ± 0.821385
   (c) 4.2 ± 0.05, 5.1 ± 0.05, 8.5143 ± 0.9709, 47.54955101 ± 17.07825836, 2479.30272768 ± 1640.69098988

INDEX